Recording Demo Tapes at Home

RELATED TITLES

Recording Demo Tapes at Home

Bruce Bartlett

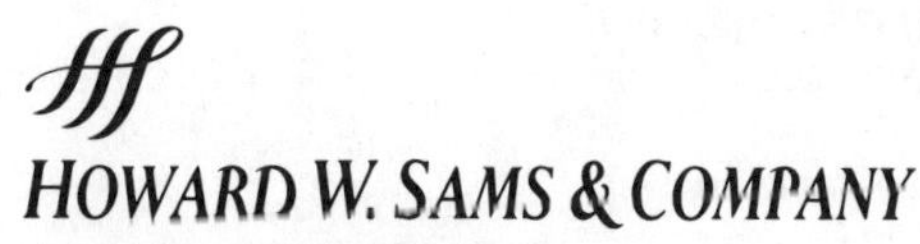

A Division of Macmillan, Inc.
4300 West 62nd Street
Indianapolis, Indiana 46268 USA

FIRST EDITION
FIRST PRINTING—1989

International Standard Book Number: 0-672-22644-8
Library of Congress Catalog Card Number: 89-60589

Acquisitions Editor: *James S. Hill*
Developing Editor: *James Rounds*
Series Editor: *John M. Woram*
Production Coordinator: *Marjorie Hopper*
Manuscript Editor: *Don MacLaren, BooksCraft, Inc., Indianapolis*
Illustrators: *Wm. D. Basham and Don Clemons*
Cover Illustrator: *April Goodman-Willy*
Compositor: *Impressions, Inc.*
Proofreader: *Gail Sax*
Keyboarder: *David Ann Gregson*
Indexer: *Don Herrington*

Printed in the United States of America

With love and appreciation to Jenny.

Overview

Contents

Preface

Home recording is hot. Many musicians are rolling their own demo tapes with portable recorder-mixers and other equipment designed for home studios.

Although home-recording equipment is becoming easier to understand and use, there's still a lot of technology to learn. So the time has come for a complete but understandable guide to home recording. This book is a nontechnical, easy-to-follow primer for musicians and audio hobbyists who want to get the best possible sound on tape for the least amount of fuss and money. It covers all you need to know—but no more—so that you can get started quickly without wading through long discussions.

Based on my work as a professional recording engineer, the book is full of tips and shortcuts for making great-sounding demo tapes at home, with up-to-date coverage of home recording equipment, setup, and procedures.

The book starts with an overview of the recording-reproduction chain to instill a system concept. Next, it covers the equipment you need and how to set it up to make a home studio. Simple 2-track recording techniques are described for those whose music can be effectively recorded this way. Further chapters explain recorder-mixer features, signal processors, and microphone techniques. With this knowledge in hand, you'll be ready to operate your equipment in a recording session in your studio. Chapter 9 tells you how, and this is followed by a chapter covering on-location recording methods.

Making a good recording requires more than choosing and operating equipment; you also need to know how to judge the recording's sound quality. Guidelines for this are given in Chapter 11. The recent technology of sampling, sequencing, and MIDI are covered next, including recording procedures for various systems from simple to complex.

The last chapter suggests some of the many uses for your demo tape. For example, you can

- document your musical ideas
- show your band how you envision your song
- send copies to friends and relatives
- enter contests
- document your musical progress
- remember your arrangements
- train new band members
- show your work to potential managers
- audition for club owners to get gigs
- obtain a recording contract
- start a professional studio production

The Appendixes present a series of experiments to train your hearing, reinforce a basic understanding of principles of sound, and list additional references.

The quality of your recording will enhance the possibilities of getting jobs or recording contracts. With a professional-sounding production, you can confidently send your tape to others who may enhance your musical career.

Acknowledgments

My deepest thanks to Jenny Bartlett for her many helpful suggestions as a consultant and editor; she made sure the book could be understood by beginners.

Thank you to John Woram for skillfully editing the book.

Thanks to Crown International for allowing me the time to work on this book.

Thank you to Larry and Elaine Zide of *db* magazine for allowing me to use material from my "Recording Techniques" series.

Finally, to all the musicians I've recorded and played with, a special thanks for teaching me indirectly about recording.

Trademarks

All terms mentioned in this book that are known to be trademarks or service marks are listed below. In addition, terms suspected of being trademarks or service marks have been appropriately capitalized. Howard W. Sams & Company cannot attest to the accuracy of this information. Use of a term in this book should not be regarded as affecting the validity of any trademark or service mark.

Aphex Aural Exciter is a registered trademark of Aphex Systems, Ltd.

Auratone is a trademark of Auratone Corp.

Commodore 64 and Commodore 128 are registered trademarks of Commodore Electronics Limited.

dbx is a registered trademark of dbx, Newton, MA, USA, Division of BSR North America, Ltd.

Dolby, Dolby A, Dolby B, Dolby C, Dolby SR, and Dolby Tone are registered trademarks of Dolby Laboratories Licensing Corporation.

IBM PC is a registered trademark of International Business Machines Corp.

Macintosh is a registered trademark of the Apple Computer Co.
Nearfield is a trademark of E.M. Long Associates.
Passport Master Tracks Pro MIDI Software is a trademark of Passport Designs, Inc.
Pressure Zone Microphone and PZM are registered trademarks of Crown International.
Radio Shack is a registered trademark of Tandy Corp.
Sony is a trademark of Sony Corp. of America.
Tascam 238 Syncaset is a registered trademark of Tascam Corp.
Variable-D is a registered trademark of Electro-Voice Inc.
Variac is a trade name of General Radio Company.
Yamaha is a registered trademark of Yamaha Electronic Corporation USA.

1 The Recording and Reproduction Chain

Recording a demo tape used to be complicated and expensive. You needed a studio full of fancy equipment, a skilled recording engineer, and many long hours to learn the technology. Today, putting your music on tape is faster, easier, and less costly—thanks to home recording equipment. The new generation of cassette recorder-mixers, special-effects devices, and MIDI sequencers enables a musician to make recordings almost like those of professional studios.

Home Studio Uses

What can you do with these creative new tools? Well, you might record a demo of yourself or your band to send to record companies. Or you might make audition tapes to help you get gigs in clubs or festivals.

With home recording equipment, you can practice your recording techniques before going into a professional studio, which saves the expense of learning the technology on studio time.

If you've written a song for your band, you can perform and record all the parts yourself: for example, rhythm guitar, bass, drums, and vocals. Then you can play this recording for the members of the band to show them how you want your song to sound.

Home recording can be educational, as well. When you perform a song, you tend to concentrate on your own instrument. But when you hear a tape playback of that song, you can listen to the song as a whole. You can hear better what works musically and what doesn't. It's less expensive to do this experimenting at home than in the studio.

Plus, there's a relaxed atmosphere at home, free of the time pressure that can hinder good playing.

Also, when you operate the equipment yourself, you avoid the tension of working with an outside engineer who may be unfamiliar with your music. You know better than anyone how you want your demo to sound, and you can make it sound just the way you like.

Why This Book Was Written

The more professional-sounding your tapes, the more jobs and contracts you're likely to receive. But making a good recording involves more than plugging a microphone into a tape deck and hitting the record button; modern recording equipment and techniques are more complicated than that. Before you can achieve a quality recording, there's a lot of equipment and terminology to understand and many techniques to learn.

This book was written to help a beginner get a firm grip on the basics of quality home recording. It separates the multitude of equipment and procedures into easily understandable parts. It lists the equipment you need, tells what it does, and suggests how to use it effectively. It also helps you choose equipment to meet your needs—anything from cheap-but-adequate to professional quality. After studying this book and practicing with actual recording equipment, you'll be making great-sounding tapes.

Overview

To begin, let's take a look at the entire process.

Musical sound starts with musicians and their instruments, goes through a series of changes and manipulations, and ends with the musicial experience in the ears and mind of the listener. The series of events and equipment that are involved in sound recording and playback is called the *recording and reproduction chain.* This chapter takes a broad view of the parts of the chain. Later chapters describe each part in detail.

The Parts of the Chain

There are many ways, from simple to complex, to record music. We'll explore the parts of the chain for three different setups.

1. Live single-point recording—recording with one or two microphones directly into a tape recorder.
2. Live mixed recording—recording with several mics into a mixer that is connected to a tape recorder.
3. Multitrack recording—recording with several mics into a mixer that is connected to a multitrack tape recorder. Each track or path on tape contains the sound of a different instrument. The multiple tape tracks are mixed after the recording session.

Live Single-Point Recording

Figure 1-1 is a diagram of the elements in this recording chain. Let's look at each element from left to right (beginning to end).

The Musical Instrument

A musical instrument is a tool to convert musical ideas and feelings into sound. Playing technique and instrument quality affect the sound the instrument produces. Sound waves travel in all directions from the instrument, and different tone qualities are heard at various positions around the instrument. The loudness of that sound is measured in decibels (dB). One dB is the smallest change in loudness that we can hear.

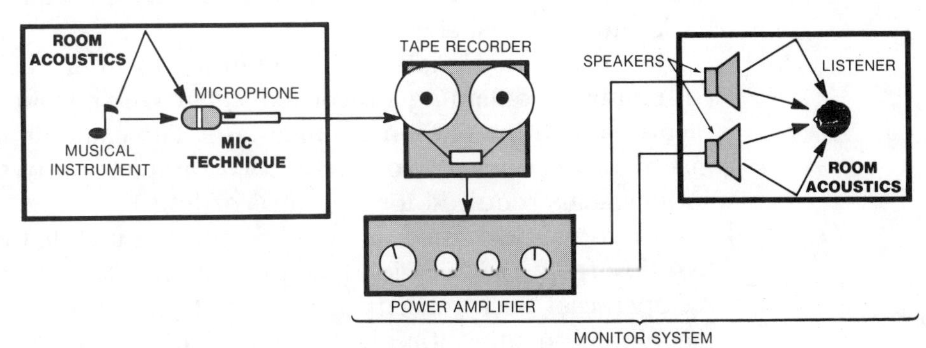

Figure 1-1. The recording-reproduction chain for recording with a microphone and a tape recorder.

Room Acoustics

After the sound waves leave the instrument, they travel through the air and reflect, or bounce, off the walls, ceiling, and floor of the recording studio. These reflections add a sense of ambience or space. If these reflections are too strong, they can make your recording sound distant and muddy. So it's common to put sound-absorbing materials on the walls, ceiling, and floor of a studio. Two examples of sound absorbers are muslin-covered, thick fiberglass insulation and thick blankets spaced out from the walls and ceiling.

In your home, the studio can be any quiet room, such as a basement, garage, or living room. The studio holds the musicians, microphones, and often the recording equipment.

The Microphone

The sound waves from the instrument travel to a microphone, which converts the sound waves into a corresponding electrical signal. At various stages in the chain the strength or level of this signal is measured in decibels.

Microphone Technique

Which microphone you choose, and where you place it, affects the tone quality and the amount of room acoustics that are recorded. Good mic technique is crucial in making a quality recording.

The Tape Recorder

Next, the electrical signal from the microphone goes into a tape recorder (such as a cassette deck) where it is converted into a magnetic signal that is stored on magnetic tape. A tape recorder acts like a time machine, storing the music in magnetic form for playback at a later date. During playback, the magnetic signals on tape are converted back into electrical signals.

The magnetic signals are stored in the form of tracks. A *track* is a path on tape containing a recorded signal. One or more tracks can be recorded side by side on a single tape. For example, a 2-track tape machine can record two tracks, for example, the two independent audio signals required for stereo recording.

A cassette deck records on a cassette, a small, flat box containing two tiny reels and ⅛″-wide tape permanently wound on those reels. An open-reel deck records on reels of tape ¼″ to 2″ wide. With this machine, you must thread the tape through the deck and onto an empty reel. Noise-reduction devices such as Dolby or dbx often are built into the recorder to reduce *tape hiss*—a rushing sound like wind in trees.

The Monitor System

To hear the signal you're recording, you need a monitor system: headphones or loudspeakers. The sound from the monitors indicates how well your recording techniques are working.

Headphones and speakers convert the electrical signal from the tape recorder back into sound. Ideally, this sound is like that produced by the original instruments. Since the electrical signal from the recorder is too weak to drive loudspeakers directly, to use them you also need a power amplifier. The room acoustics affect the sound leaving the speakers and reaching the ears of the listener. Here we arrive at the end of the chain.

Live Mixed Recording

Let's move on to a more complex way to record, which is illustrated in Figure 1-2. It starts with the musical instruments, room acoustics, microphone techniques, and microphones already described. But now there's a microphone for each musician or singer. All these microphones plug into a mixer, which blends, or mixes, all the microphone signals into one.

The Mixer or Mixing Console

The mixer has a volume control for each microphone so the volume of each instrument's signal can be adjusted to make a pleasing loudness balance. For example, if the guitar is too quiet relative to the voice, you turn up the volume control for the guitar microphone until it blends well with the voice.

Why record this way? Because it's easier to adjust volume controls to achieve a good balance than to adjust the musicians' positions.

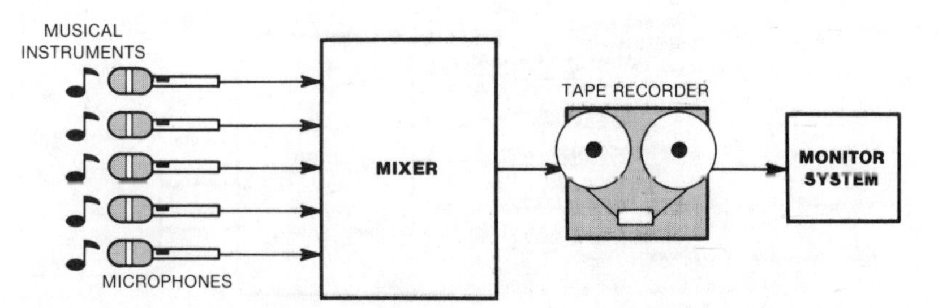

Figure 1-2. The recording-reproduction chain for recording with multiplo microphones and a mixer.

What's more, since each microphone is close to its instrument, it mainly picks up the sound of that instrument and very little room acoustics. The result is a clearer sound.

The output signal of the mixer is recorded with a tape recorder, and you listen to the signal with the monitor system already described.

Many mixers let you control other aspects of sound besides volume. You can control tone quality (bass and treble), stereo position (left, right, or center), and special effects (such as artificial reverberation, which sounds like room acoustics).

Multitrack Recording

The only problem with the previous setup is that you have to mix the sound of the musicians as they're playing. If you make a mistake while mixing—say, one instrument is too quiet—the musicians have to play the song again until you get the balance right. Wouldn't it be great to record the signal of each microphone independently, then mix these recorded signals after the performance is done? Well, you can.

The Multitrack Recorder

A multitrack recorder records from 4 to 32 tracks side by side on a tape. You can record a different instrument on each track (as diagrammed in Figure 1-3). Or, if you have more instruments than tracks, you can record different groups of instruments on each track.

The Mixing Console

During multitrack recording the mixing console is mainly used to amplify the weak microphone signals to a level suitable for the tape recorder and to assign each microphone signal to the desired track.

After the recording session, you play all the tracks through the mixer to mix them with a pleasing balance. That is, you mix the re-

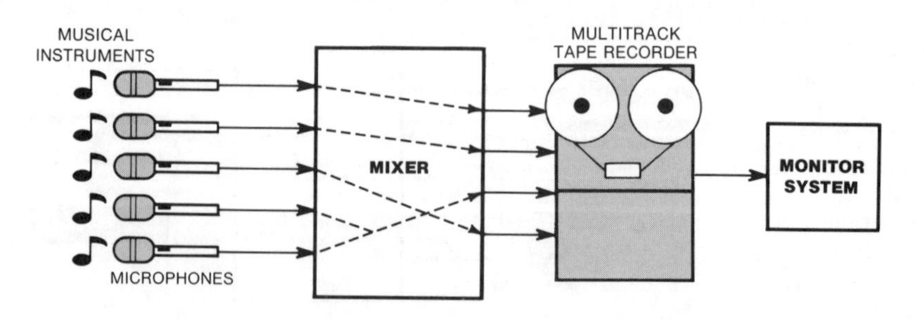

Figure 1-3. Multitrack recording.

corded signals from the tape tracks, rather than the live signals from the microphones.

You play back the multitrack tape of the performance several times until the mix is perfected. Your final mix is recorded on a 2-track stereo tape recorder (open-reel or cassette) as shown in Figure 1-4, and the resulting 2-track tape is the final product.

Overdubbing

There's another benefit of multitrack recording. You can record a few instruments, and then go back later to record more instruments on unused tracks. This process is called *overdubbing.* If you play several instruments, you can record yourself playing one instrument at a time. With each additional instrument, you listen to the previously recorded tracks (to keep your place in the song) and play along with them.

The Cue System

If several musicians are overdubbing at once, you need a system that enables them to hear each other, and previously recorded material, through headphones. Many mixing consoles include a *cue mixer* built in, which is used to create a cue mix heard over headphones. The signal from the cue mixer is strengthened by a power amplifier before it goes to a multiple-connector box which the headphones plug into.

Effects

You can connect external devices called *signal processors* to your mixer in order to produce special effects to enhance the sound quality. These effects include reverberation, echo, and compression.

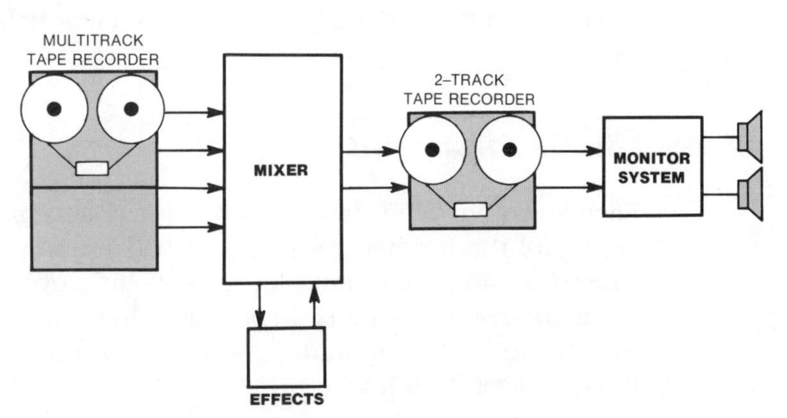

Figure 1-4. Multitrack mixdown.

Summary

Let's review all the changes music goes through from the start of the recording-reproduction chain to the end:

1. The musical instrument converts motion into sound waves.
2. The resulting sound waves are modified by room reflections (studio acoustics).
3. At the microphone, the modified sound waves are converted into electricity (the signal).
4. The signal from the microphone is affected by microphone selection and placement (microphone technique).
5. All the microphone signals are controlled and modified by the mixing console and signal processors.
6. The modified electrical signal is recorded—changed into a magnetic signal for storage.
7. The magnetic signal is played back—changed into an electrical signal.
8. The electrical signal is amplified and changed back into sound waves by the monitor amplifier and speakers.
9. The loudspeaker sound is modified by sound reflections in the listening room.
10. The sound strikes the listener's ears and is heard as music.

While all this is going on, if a cue system is used, the musicians can be listening to each other and the tape through headphones.

The end product of the recording end of the chain is the master tape. Additional links in the chain include tape copies and the listener's playback system.

Every Link Is Important

Each link of the recording-reproduction chain contributes to the sound quality of the finished recording. A bad-sounding master tape can be caused by any weak link: low-quality microphones, bad mic placement, improperly set mixer controls, and so on. A good-sounding tape results when you optimize every part of the chain. This book tells how to reach that goal.

2 Equipping Your Home Recording System

It's every musician's dream. You want to set up a home recording system—one with good-quality sound, yet affordable. With today's equipment, you can do just that. We can divide the available equipment into five levels based on price and sophistication.

1. A high-quality stereo cassette recorder with two mics costs $135 and up. This is not suitable for recording amplified groups, such as rock bands, but it works very well for a soloist or acoustic groups, such as vocal quartets, solo vocalists or pianists, folk groups, symphonic bands, or orchestras. Depending on your room acoustics and the balance you achieve, the quality can be good enough to make a demo tape to send to a record company. Some commercial recordings have been made this way. It's certainly adequate for an audition tape for club owners and music publishers.
2. A setup with multiple microphones, a mixer, and a stereo cassette recorder costs $500 to $1,000, depending on the sophistication of the mixer. To add optional effects costs $200 or more. You may already have an adequate mixer and microphones that you use for sound reinforcement. Sometimes these can be used for recording as well. This arrangement is suitable for audition, demo, or album recordings of a self-accompanied singer or of small vocal, folk, jazz, and bluegrass groups. It also can be used for audition tapes of rock groups.

   A drawback of this system is that you have to mix the music as it is performed, so the mix may not be optimum. A multitrack recorder lets you record the instruments and vocals on separate tracks and mix them after the recording session. It also lets you overdub. A tape made with overdubs is potentially clearer than a tape made with all the instruments and vocals recorded at once.

3. A good-quality 4-track recorder-mixer setup (including microphones and mic stands) costs about $700. To add optional effects costs $200 or more. This system can be used to document your ideas, to work out musical arrangements, or to play your song ideas to your group. It's good enough to make audition tapes for club owners and music publishers, but not good enough to make a demo tape to send to a record company. If you want to learn the basics of multitrack recording without spending a lot, this is the way to go.
4. A high-quality 4-track recorder-mixer system costs about $2,200. To add optional effects costs $200 or more. This setup is good enough to make demos to send to record companies or to make audition tapes for club owners and music publishers, but it is not generally considered good enough to make commercial recordings.
5. A high-quality 8-track outfit can be assembled for under $6,000. To add optional effects costs $200 or more. This system is good enough to make commercial recordings or high-quality demos of small groups.

With any of these systems, you can save money if you already have some of the equipment. Try it out; if it sounds good to you, you're all set.

Table 2-1 suggests which recording systems are best for various musical sources. Live-to-2-track systems are adequate for many groups and generally cost less than multitrack systems. Multitrack systems let you perfect the mix after the session and add overdubs, which results in clearer sound in some ases. Audition tapes are of good quality and can be used to get gigs. Demo tapes are of the best quality and can be sent to record companies.

The remainder of this chapter describes in detail the equipment in each type of system.

Table 2-1. Suggested recording systems for various musical sources.

Musical Source	Live to 2- Track		Multitrack	
	Audition	**Demo**	**Audition**	**Demo**
Classical Soloist	A	A		
Singer plus piano or guitar	A	B	C	D
Vocal quartet with good blend and good acoustics	A	A		
Vocal quartet without good blend or good acoustics	B	B	C	D
Folk group/ethnic band with good blend and good acoustics	A	A	C	D,E
Folk group/ethnic band without good blend or good acoustics	A	B	C	D,E
Orchestra, symphonic band, pipe organ, string quartet	A	A		
Bluegrass band	B	B	D	E
Jazz group	B	B	C	D,E
Rock/pop/country/soul group	B		C	D,E

Key:
A: two mics and a cassette deck
B: several mics, mixer, cassette deck
C: budget recorder-mixer, mics
D: high-quality recorder-mixer, mics
E: 8-track recorder, mixer, mics

The Stereo Cassette System

As mentioned before, this simple setup is used for recording a soloist or a traditional acoustic ensemble. All you need is:

1 quality cassette deck with dbx or Dolby noise reduction	$75 and up
2 identical mics or a stereo mic	$45 and up
1 or 2 mic stands (these are unnecessary if you use miniature mics)	$15 each
Total	$135 and up

Cassette Deck

The deck should include dbx or Dolby noise reduction to reduce tape hiss. Tips on choosing a cassette deck are given in Chapter 4.

Microphones

You'll need two microphones of the same model number. The cardioid condenser type gives the clearest sound. Or you can use a stereo microphone, which combines two mics in a single housing and can be mounted on a single stand.

Good mics are essential for quality sound. Generally, you get what you pay for, and while some people are happy to get any sound on tape, others will settle for nothing less than professional sound quality. There are big differences in fidelity among the various types of mics. You can use hand-me-down mics if they sound good enough.

Unlisted Equipment

You need blank cassette tape. Use the tape suggested by the cassette-deck manufacturer. Brand-name metal or chromium dioxide tape is recommended for best sound quality.

Good headphones for monitoring are available for about $50 and up. You can also use a quality home stereo system (speakers, etc.) to listen to the playback.

The Budget Multitrack System

A budget system suitable for documenting your ideas or making audition tapes might include the following equipment:

4-track recorder-mixer	$449
2 mics ($100 each)	200
2 mic stands and booms	55
Total	$704

Recorder-Mixer (Portable Studio or Ministudio)

This is a small, portable unit combining a mixer with a multitrack recorder. In a budget system, the recorder-mixer is a 4-track cassette recorder with a built-in two-input mixer. You can record one or two tracks at a time, building up to four tracks for later mixing down to 2-track stereo. For example, you might record a keyboard part on one

Bookmark

HOWARD W. SAMS & COMPANY

DEAR VALUED CUSTOMER:

Howard W. Sams & Company is dedicated to bringing you timely and authoritative books for your personal and professional library. Our goal is to provide you with excellent technical books written by the most qualified authors. You can assist us in this endeavor by checking the box next to your particular areas of interest.

We appreciate your comments and will use the information to provide you with a more comprehensive selection of titles.

Thank you,

Vice President, Book Publishing
Howard W. Sams & Company

COMPUTER TITLES:

Hardware
- ☐ Apple 140
- ☐ Macintosh 101
- ☐ Commodore 110
- ☐ IBM & Compatibles 114

Business Applications
- ☐ Word Processing J01
- ☐ Data Base J04
- ☐ Spreadsheets J02

Operating Systems
- ☐ MS-DOS K05
- ☐ OS/2 K10
- ☐ CP/M K01
- ☐ UNIX K03

Programming Languages
- ☐ C L03
- ☐ Pascal L05
- ☐ Prolog L12
- ☐ Assembly L01
- ☐ BASIC L02
- ☐ HyperTalk L14

Troubleshooting & Repair
- ☐ Computers S05
- ☐ Peripherals S10

Other
- ☐ Communications/Networking M03
- ☐ AI/Expert Systems T18

ELECTRONICS TITLES:
- ☐ Amateur Radio T01
- ☐ Audio T03
- ☐ Basic Electronics T20
- ☐ Basic Electricity T21
- ☐ Electronics Design T12
- ☐ Electronics Projects T04
- ☐ Satellites T09
- ☐ Instrumentation T05
- ☐ Digital Electronics T11

Troubleshooting & Repair
- ☐ Audio S11
- ☐ Television S04
- ☐ VCR S01
- ☐ Compact Disc S02
- ☐ Automotive S06
- ☐ Microwave Oven S03

Other interests or comments: ____________________

Name ____________________

Title ____________________

Company ____________________

Address ____________________

City ____________________

State/Zip ____________________

Daytime Telephone No. ____________________

A Division of Macmillan, Inc.
4300 West 62nd Street Indianapolis, Indiana 46268

22644

track and then add bass, drums, and vocal on the remaining tracks. With most units of this price range, no control of special effects is available. Features of recorders and mixers are described later in this chapter.

Blank cassette tape is needed for the recorder-mixer. Metal or chromium dioxide tape is recommended for best sound quality.

Microphones

Two microphones costing at least $100 each are recommended. Although $100 may seem like a lot of money for a microphone, you can't skimp here and expect to get quality sound. Any distortion or coloration in the microphone may be difficult or impossible to remove from the tape later on. You may be able to borrow some good microphones or use the ones you normally use in performance with your PA system. Your ears will tell you if the fidelity is adequate for your purpose. A pickup built into a stringed instrument usually doesn't sound as good as a microphone.

Microphone Types

Let's take a minute to explain microphone types and characteristics. There are three basic types of microphones for recording: condenser, moving coil, and ribbon. Each type has a different way of converting sound into electricity:

- In a *condenser* (or *electret condenser*) microphone, a diaphragm and an adjacent metallic disk (backplate) are permanently charged with static electricity. When sound waves vibrate the diaphragm, the microphone produces an electrical signal similar to the incoming sound wave.
- In a *moving-coil* (or *dynamic*) microphone, a coil of wire attached to a diaphragm is suspended in a magnetic field. When sound waves vibrate the diaphragm, the microphone generates an electrical signal similar to the incoming sound wave.
- In a *ribbon* microphone, a thin metal foil, or ribbon, is suspended in a magnetic field and generates a signal when vibrated by sound waves.

The condenser mic is commonly used on cymbals, acoustic instruments, and studio vocals. The moving-coil unit is typically used on drums and electric-guitar amplifiers. The ribbon microphone usually provides a warm, smooth sound, but it is delicate and so should not be used inside a kick drum.

The condenser mic requires a battery or phantom power to operate. A phantom power supply is a circuit that supplies dc powering to condenser microphones, using the same cable wires that the audio signal uses. Phantom power is built into the more sophisticated mixing consoles; the mic simply plugs into the mixer to receive power.

Microphone Polar Patterns

Microphones are also classified by their polar patterns or directional pickup patterns: omnidirectional, unidirectional (cardioid), or bidirectional. An *omnidirectional* (*omni*) microphone picks up sound equally well from all directions. A *cardioid* microphone rejects sound from the rear. It also rejects room acoustics and sound from other instruments, which results in a tighter, clearer sound. A *bidirectional* microphone picks up from the front and rear but rejects sound approaching the sides of the microphone.

What type of microphones is most useful? Since room acoustics and unwanted sounds are problems in home studios, a cardioid microphone is a good choice. If you get a cardioid condenser microphone (such as that shown in Figure 2-1) and a cardioid dynamic microphone (such as that shown in Figure 2-2), you can faithfully pick up a wide variety of instruments.

Three types of special-purpose microphones are:

- The miniature condenser microphone, which attaches to instruments. A drum set might be recorded with two or three miniature omni condenser mics.

Figure 2-1. Audio-Technica AT813, an example of a cardioid condenser microphone. *(Courtesy of Audio-Technica U.S., Inc.)*

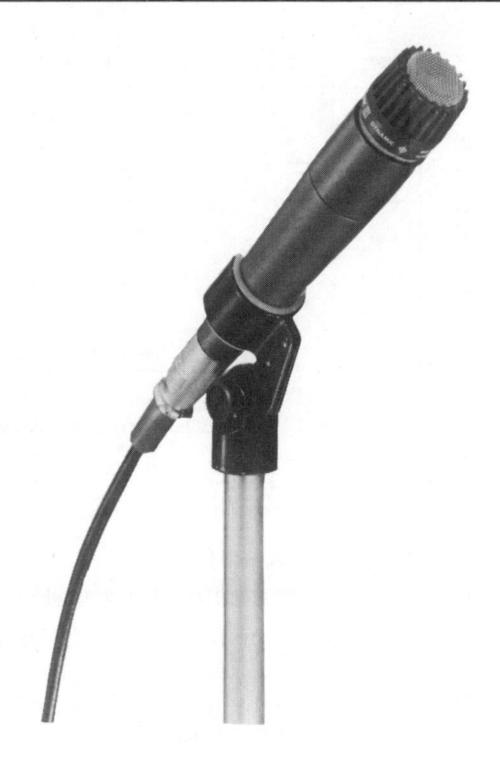

Figure 2-2. Shure SM-57, an example of a cardioid dynamic microphone. *(Courtesy of Shure Brothers, Inc.)*

- The boundary microphone (such as the Crown PZM series), which is used on such surfaces as walls, panels, and piano lids.
- The stereo microphone, which includes two directional mic capsules angled apart in a single housing and is used for recording in stereo.

Microphone Frequency Response

An important characteristic of a microphone is its frequency response. To understand what this is, first we need to understand "frequency." Low-pitched sounds have a low frequency or rate of vibration; high-pitched sounds have a high frequency. Frequency is measured in cycles per second, or hertz (Hz). One thousand hertz equal one kilohertz (kHz).

The frequency response of a microphone is the range of frequencies it will reproduce at an equal level (within a tolerance, such as ±3 dB). It is specified from the lowest frequency to the highest that the microphone will pick up within that tolerance, as, for example, 50 Hz to 14 kHz ±3 dB (shown in Figure 2-3). In general, the lower the lower figure and the higher the upper figure, the better the fidelity.

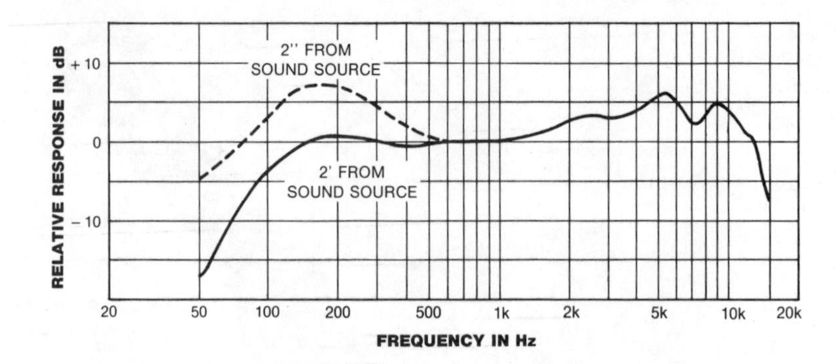

Figure 2-3. Example of a microphone frequency response.

Listed here are the frequency responses needed for fidelity with various instruments:

Bass and kick drum: response down to 40 Hz or lower

Cymbals: response up to 15 kHz or higher

Most other instruments: 80 Hz to 12 kHz or higher

Symphonic band or orchestra: 40 Hz to 15 kHz or higher

If the microphone is intended to record cymbals, acoustic instruments, or vocals, the response should be flat (±3 dB or less). If the mic is meant to record drums or electric-guitar amplifiers, the response can have a "presence peak" rising around 5 kHz. This is a suggestion, not a rule. If a mic sounds good to you on a particular instrument or voice, use it.

Microphone Impedance

Impedance is an electrical characteristic of a microphone, measured in ohms (Ω). The microphone impedance, which can be found on the mic's spec sheet, should be low (150 to 600 Ω). If you use a high-impedance mic with a cable longer than about 10 feet ('), the cable will pick up hum and the sound will be dull. A low-impedance microphone allows long cable runs without these problems. Don't bother with high-impedance mics. The symbol for impedance is Z.

Balanced vs. Unbalanced Microphones

If your mixer input connector has three little holes arranged in a triangle, it's a balanced input. This is called an XLR or three-pin pro audio connector. A microphone plugged into this kind of input should have a balanced output. If your mixer input connector is a single ¼"-diameter hole (like a phone jack), that's an unbalanced input. With this input, you can use a microphone with either a balanced or unbalanced output as long as you put a matching connector on the mic cable.

Unlisted Items

These necessary items are not included on the list because you probably already have them:

- Stereo cassette deck—use this to record your final stereo mix of the four tape tracks. You'll still have your 4-track tape, which could be remixed at a later date.
- Monitor system—this lets you hear what you're recording and mixing. It includes either headphones, or a stereo power amplifier and accurate loudspeakers. A good home stereo system or quality headphones are adequate for monitoring in budget recording systems. A typical studio-quality speaker suitable for home monitoring use is the Yamaha NS-10M. Also available are powered minispeakers with a built-in amplifier. They lack deep bass but take up little room.

The Recorder-Mixer Demo System

This system uses a full-function mixer with a built-in high-quality 4-track cassette recorder. The recorder-mixer can record up to four tracks at a time and control special effects such as artificial reverberation. A system using such a recorder-mixer is good enough to make demo recordings. In some cases, notably with folk music, it can be used to record commercial tapes and albums.

Typical equipment includes:

4-track recorder-mixer	$1,300
4 mics ($150 each)	600
4 mic stands and booms	110
4 cue headphones (for overdubs only)	160
1 direct box	50
Total	$2,220

Direct Box

This is a transformer or circuit that adapts the output of an electric instrument to the input of a mixer. A direct box allows you to plug an electric guitar, electric bass, synthesizer, or electric piano directly into the mixer for a cleaner sound. All these instruments produce an electrical signal that can feed a mixer directly. Note that a direct box picks up a very clean sound, which might be occasionally undesirable for electric guitar. If you want to pick up the distortion of the guitar amp, use a microphone instead.

Note: Some inexpensive mixers have phone jacks for mic inputs. If you use a short cable to minimize hum, you can plug an electric instrument directly into such an input without using a direct box.

You can buy a direct box for as little as $50. Or you can assemble one for $15 with a phone-plug Y-adapter and a microphone impedance-matching transformer (available from your local electronics dealer). This assembly is shown in Figure 2-4.

There's a low-cost alternative to a direct box. You, or a friend interested in electronics, can solder together some direct-connection cables as shown in Figure 2-5. This cable reduces the amplifier's output signal to a lower voltage suitable for a mixer mic input. It also reduces the treble to simulate what a guitar-amp loudspeaker does. Figure 2-6 shows a direct-connection cable for synthesizer, drum machine, or electric piano. It does not reduce the treble.

Compared to a microphone, the direct-connection cable costs very little, doesn't pick up sounds from other instruments, and doesn't pick up room acoustics. To use the cable, plug the phone plug into the guitar-amplifier external-speaker jack (or the output jack of a synthesizer, drum machine, or electric piano) and plug the other end into a mixer mic input. Flip the guitar-amp ground switch to the position that produces the least hum.

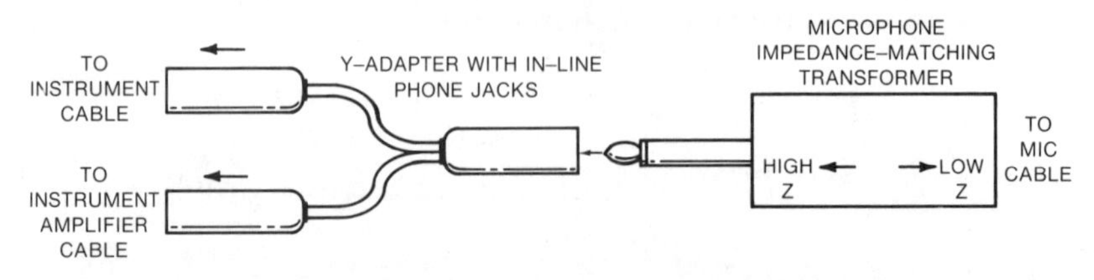

Figure 2-4. A direct box assembled from a phone-plug Y-adapter and a mic impedance-matching transformer.

Figure 2-5. Direct connection cable for an electric guitar amplifier. *(Courtesy of Steve Julstrom)*

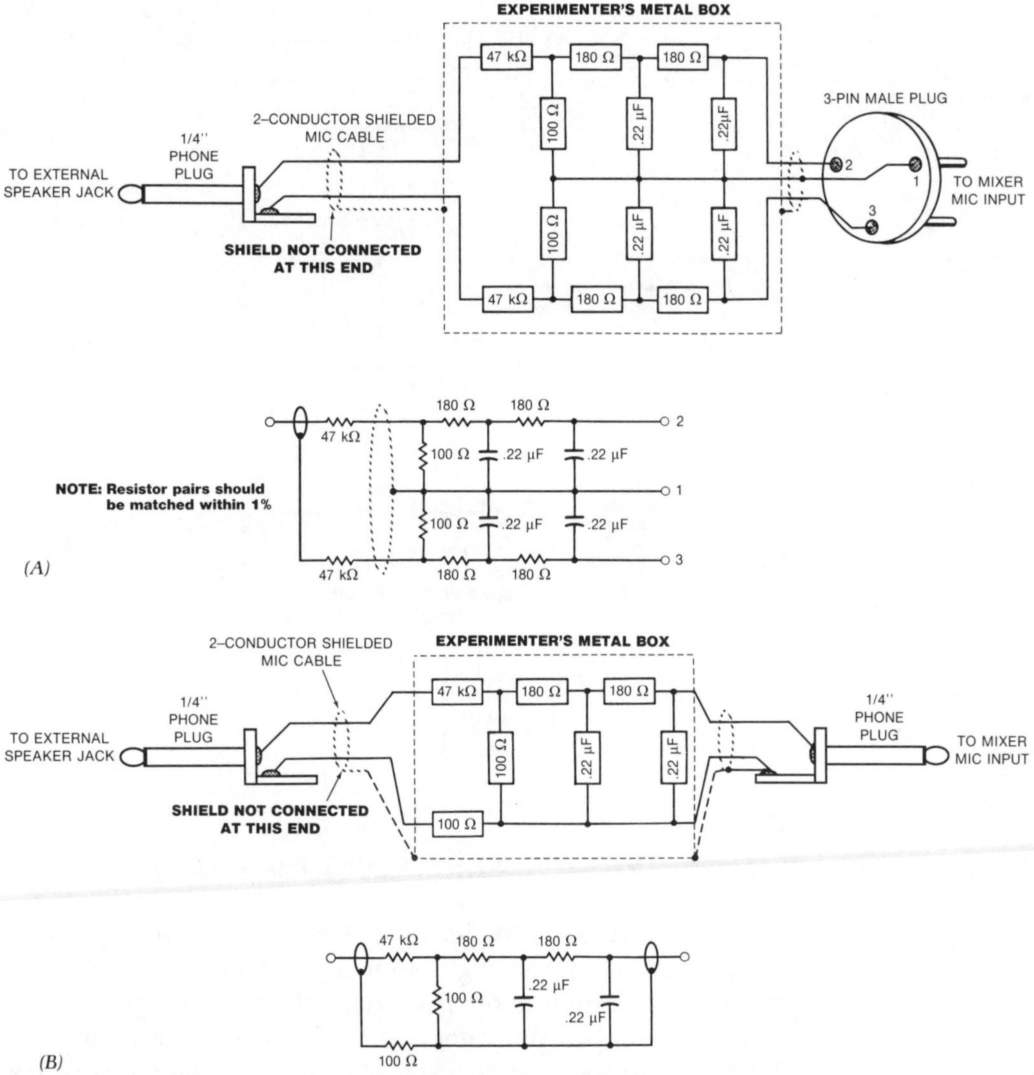

NOTE: Resistor pairs should be matched within 1%

(A)

(B)

Cue Headphones (for Overdubs)

With these headphones, musicians listen to previously recorded tracks and record a new part along with those tracks. The cue headphones plug into a box containing several headphone jacks wired in parallel.

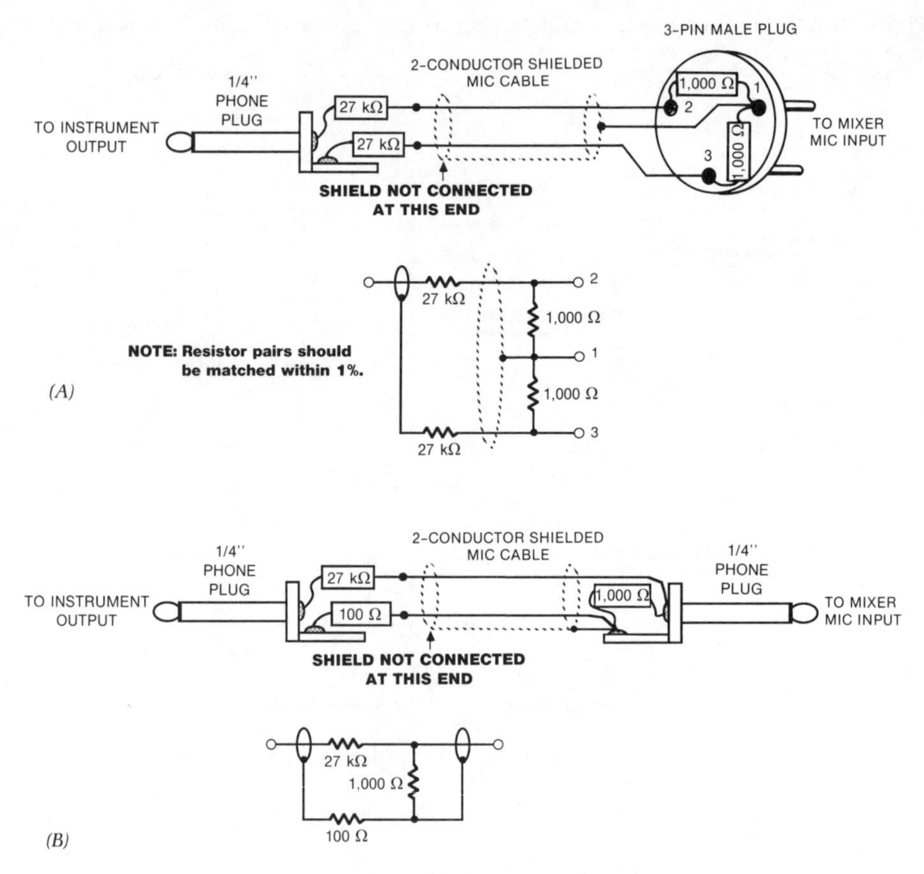

Figure 2-6. Direct-connection cable for a synthesizer, drum machine, or electric piano.

Recorder-Mixer (Portable Studio or Ministudio)

The more advanced recorder-mixers record up to four tracks at a time, have from four to eight inputs, and allow control of special effects. They are good enough to make demo recordings.

While recorder-mixers offer good performance for their price, they can't compete in sound with a separate mixer and open-reel recorder. Cassette recorders generally have poorer fidelity than open-reel recorders. Still, you can use recorder-mixers whenever the mood strikes; they're affordable, the tapes are cheaper and easier to handle, and you can communicate your music with them.

Prices of cassette recorder-mixers range from $449 for simple personal studios, to $895 for intermediate units, and as much as $1,300 or more for top-quality units.

Recorder-mixers currently on the market are made by Fostex, Tascam, Audio-Technica, AMR, Yamaha, Cutek, Vesta Fire, Ross, and Clarion. Three examples are shown in Figures 2-7, 2-8, and 2-9. Features of recorder-mixers are described in detail in Chapter 5.

The 8-Track System

This type of system has an 8-track recorder and a separate mixer. It can be used professionally. A list of equipment includes:

8-in, 4-out mixer	$1,300
8-track recorder	1,900
2-track open-reel recorder	775
6 mics ($200 each)	1,200
6 mic stands and booms	165
1 mic snake	200
4 cue headphones	160
Tape	75
Total	$5,775

8-Track Recorder

An 8-track recorder (such as those shown in Figures 2-10, 2-11, and 2-12) lets you record up to eight independent tracks, each containing

Figure 2-7. Fostex Model 260 Multitracker. *(Courtesy of Fostex Corporation of America)*

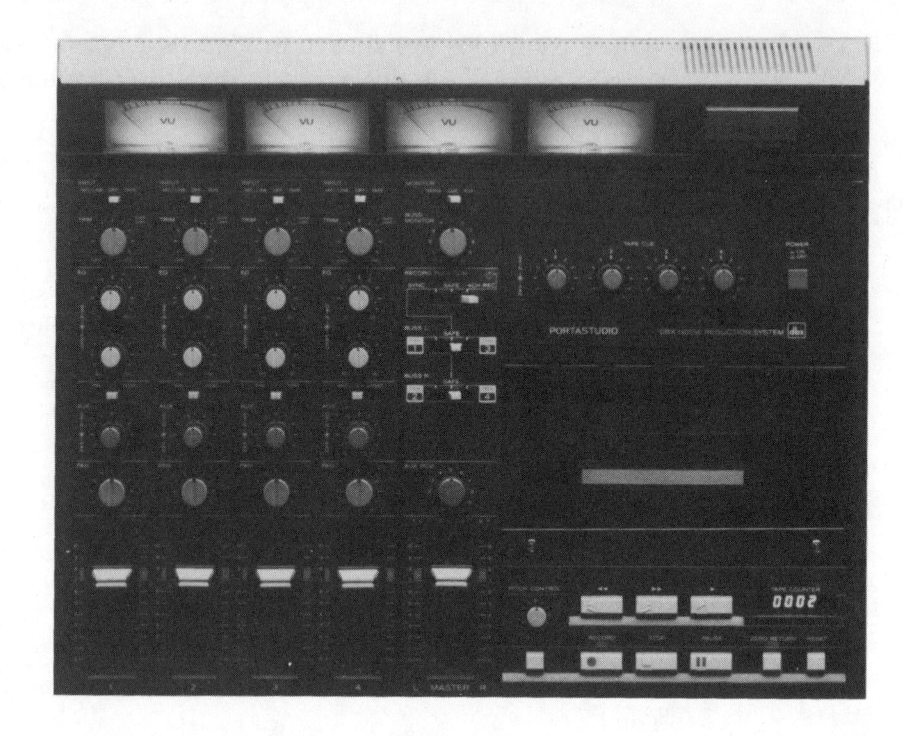

Figure 2-8. Tascam Model 244 Portastudio. *(Courtesy of Tascam Corporation of America)*

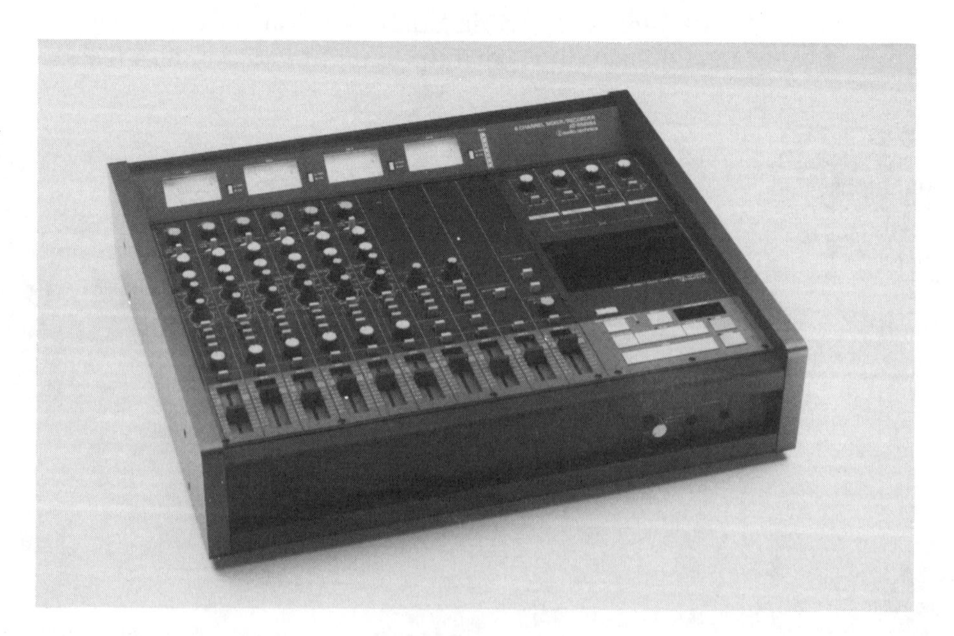

Figure 2-9. Audio-Technica Model AT-RMX64. *(Courtesy of Audio-Technica U.S. Inc.)*

Figure 2-10. Tascam 38 8-track recorder. *(Courtesy of Tascam Professional Products)*

the sound of one or more instruments. An 8-track unit, including built-in noise reduction, for home-studio use can be bought for about $2,000. A 16-track recorder with built-in noise reduction can be bought for less than $6,000. The sound quality of the open-reel decks is good enough for use in making records. An 8-track machine is much more convenient to use than a 4-track machine. Often, when you record a band on a 4-track unit, you must combine several instruments on each track. As a result, you can't adjust the level, tone, or effects of each instrument independently within a track once it is recorded. Eight tracks usually are enough to record each instrument on its own track, so you can control the sound of each instrument individually.

Microphones

Do you need eight mics for eight tracks? Not necessarily. The number of mics you need depends on your instrumentation. You might want to use eight mics on the drums *alone*. You can mix the eight drum mics to one or two tracks. Or you might want to use one mic for vocals and record drum machines and synthesizers directly. You can use one microphone for several different instruments if you overdub them.

Figure 2-11. Fostex Model-80 8-track recorder. *(Courtesy of Fostex Corporation of America)*

Figure 2-12. Tascam 238 Syncaset 8-track cassette recorder. *(Courtesy of Tascam Professional Products)*

2-Track Open-Reel Recorder

This is used to record stereo mixes of the eight tracks you recorded on the 8-track machine. An open-reel recorder is used rather than a cassette deck because the open-reel format reproduces sound with higher quality, permits tape editing, and can be used to cut records. Figure 2-13 shows an example of such a recorder.

The 8-In 4-Out Mixer

Although an eight-output mixer is commonly used with an 8-track recorder, a less expensive alternative is a 4-output mixer (such as those shown in Figures 2-14 and 2-15). You can use it with an 8-track recorder by recording four tracks at a time or by using the direct-out jacks in the mixer. If the mixer has eight inputs, you can record eight tracks simultaneously (one instrument per track) from the direct outputs.

Combination open-reel recorder-mixers are available in 8-track format (such as that shown in Figure 2-16) or 14-track format (such as that shown in Figure 2-17). This unusual format has 12 audio tracks, one control track, and one sync track.

Figure 2-13. Fostex Model-20 2-Track Open-Reel Recorder. *(Courtesy of Fostex Corporation of America)*

Figure 2-14. Tascam M-308 Mixing Console. *(Courtesy of Tascam Professional Products)*

Figure 2-15. Fostex Model 450 Mixing Console. *(Courtesy of Fostex Corporation of America)*

Microphone Snake

If your band has an engineer who doesn't play in the band, the engineer can operate the equipment and monitor the recording in a control room that is separate from the studio. Long microphone cables can be used to carry the mic signals from the studio to the control room, but it's messy and time-consuming to run all these cables. Instead, you can use a microphone snake, which has several mic con-

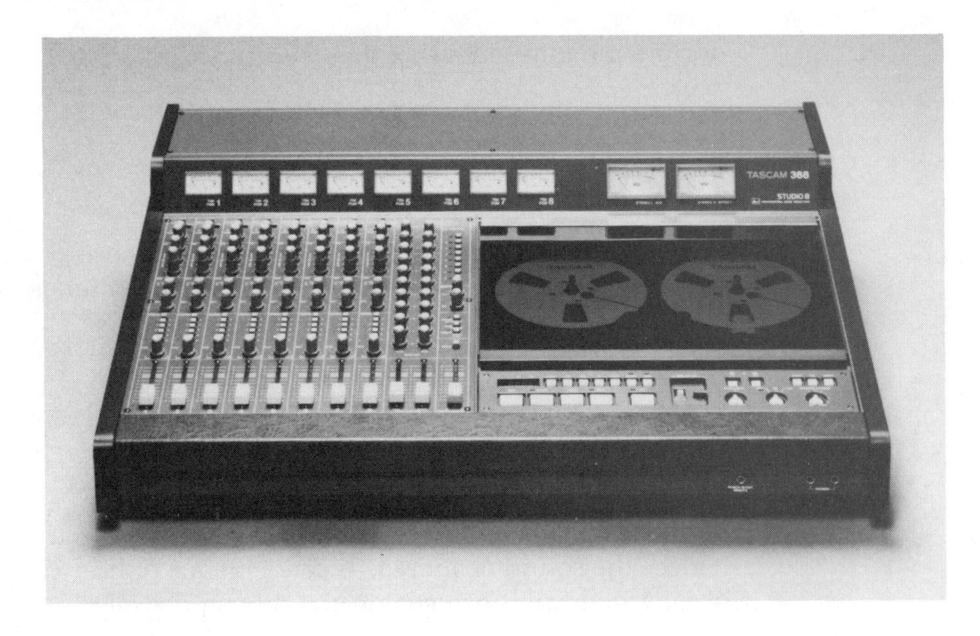

Figure 2-16. Tascam 388 Studio 8, an 8-track recorder-mixer. *(Courtesy of Tascam Professional Products)*

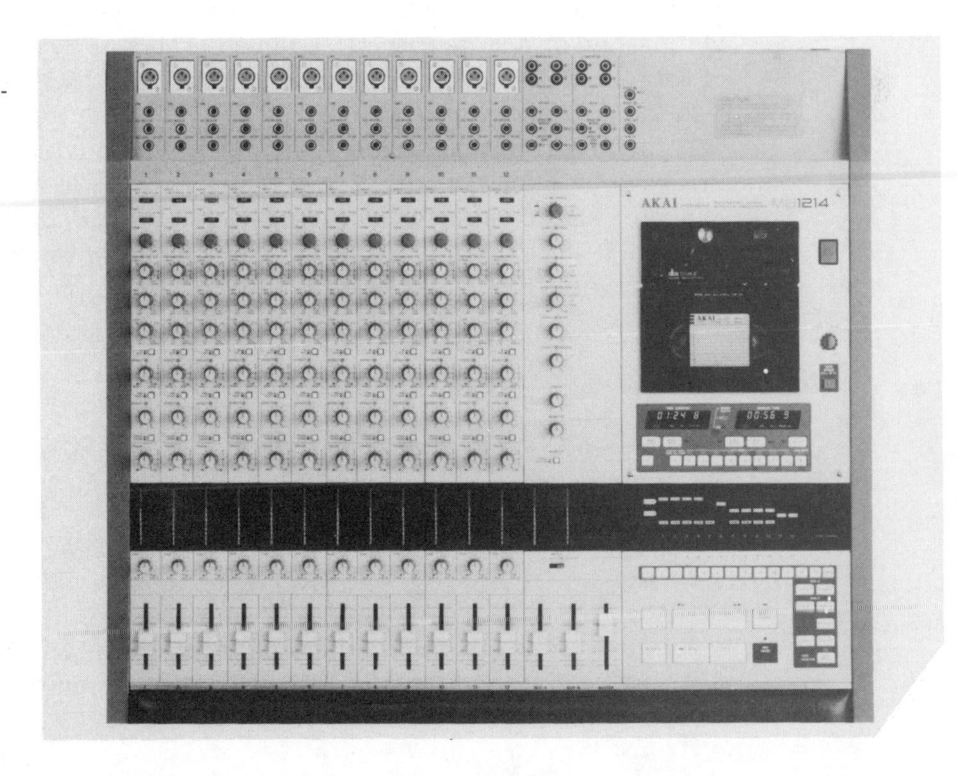

Figure 2-17. Akai MG1214 12-channel mixer-14-track recorder. *(Courtesy of Akai America Limited)*

nectors mounted in a box that is attached to a thick, multiconductor cable.

Blank Tape

Don't forget this expense! Buy the best low-noise, high-output tape you can afford. If possible, use the brand recommended by the recorder manufacturer.

Optional Extras

The following equipment is optional. Each piece of equipment enhances the sound or increases convenience of use. This equipment is usually *outboard*—external to the mixing console.

Reverberation Unit

Reverberation is the sound you hear just after you shout in an empty gymnasium or a large cathedral. Reverberation is a continuous decay of sound (''Hello-o-o-o-o-o''), while an echo is a discrete repetition of a sound (''Hello hello hello hello'').

Most recordings are made in a ''dead'' studio which has little natural reverberation. To add a sense of spaciousness to recordings made in such a studio, an artificial reverberation device (*digital reverb*) can be used. It electronically simulates room acoustics or ambience. Good digital reverbs are available for about $200 and up.

Digital Delay

This provides many effects, such as echo, chorus, doubling, and flanging, which add some pizazz to the recorded sound.

Compressor

A compressor keeps the loudness of vocals (or any instrument) constant, acting like an automatic volume control. Home-studio units start around $125.

Rack-Mounted Patch Panel

All this optional (outboard) equipment can be mounted in a rack, a wooden or metal enclosure with mounting holes for equipment. You could also install a patch panel or patch bay: an array of connectors that are wired to equipment inputs and outputs. A rack with a patch panel is shown in Figure 2-18. Using a patch panel and patch cords, you can change equipment connections easily. Also you can bypass or patch around defective equipment. Figure 2-19 shows the typical usage of various patch-panel jacks.

Acoustic Treatment

This is material placed in the studio to absorb excessive sound reflections. It results in a cleaner recorded sound by controlling room reverberation.

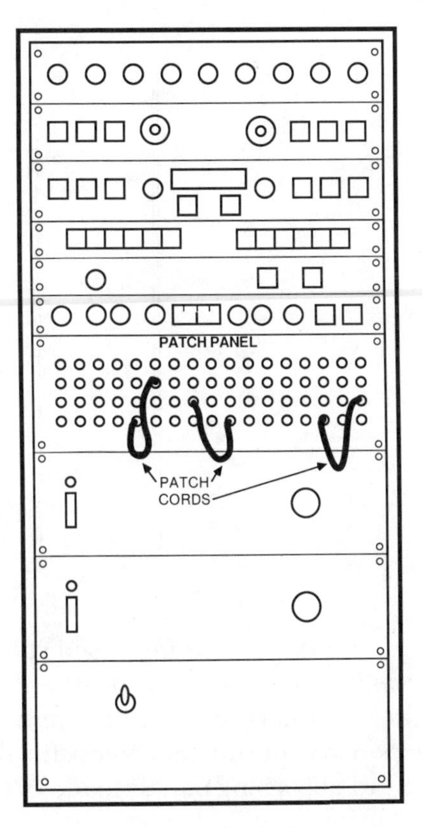

Figure 2-18.
A rack with a patch panel.

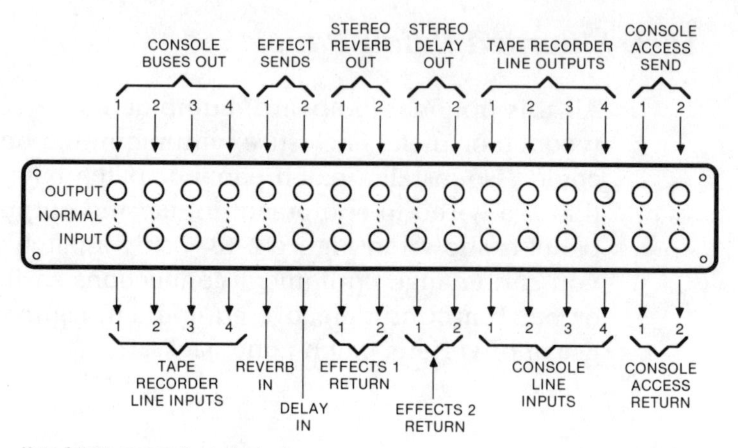

Figure 2-19. Input/output panel jacks.

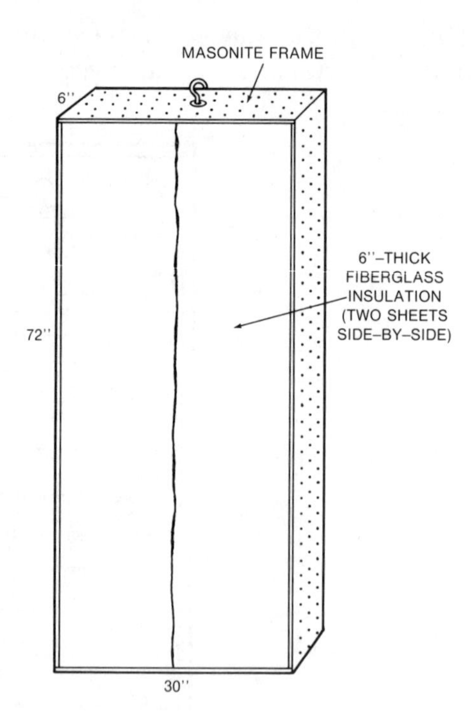

Figure 2-20. A sound-absorbing panel.

For budget or improvised studios, acoustic treatment is necessarily limited. Try surrounding the instrument and its microphone with thick blankets or sleeping bags hung a few feet away. For some inexpensive permanent acoustic treatment, carpet the floor and add sound-absorbing panels made in the following way (refer also to Figure 2-20):

1. Build a 30″ × 70″ × 6″ rectangular frame of perforated masonite.
2. Place two strips of 6″-thick fiberglass insulation with the paper or foil cover removed, side by side inside the frame.
3. Cover the frame with muslin or burlap.
4. Screw a hook into the top for hanging, or balance the panel against the wall.

Start with about eight of these panels. Place them near the instrument you're recording or space them evenly around the walls. If necessary, add panels until your recordings sound reasonably dry (free of audible room reverberation).

Conclusion

As we've seen, putting together a high-quality home recording system needn't cost much, and better equipment is continually being produced at lower prices. That dream of making your own demo tape is within reach.

3 Setting Up the System

Okay! You've got your recording gear, so it's time to set it up and hook it together. We'll work on room acoustics, equipment placement, and connections.

One Room or Two?

Some bands play their music and operate the recording equipment in the same room. Others use a separate control room to hold the recording equipment. Which is better? Let's consider the options.

If you're playing in a band and operating the recording equipment at the same time, the equipment will have to be in the room in which you're playing. In that case, though, it might be hard for you to hear over headphones what kind of sound the microphones are picking up because the band's sound drowns out the sound heard over headphones. To solve this problem, make some trial recordings and play them back to check the sound quality. Make the necessary adjustments (described in Chapter 9) and rerecord.

If you're recording yourself playing an electric instrument that is plugged directly into the recording equipment, then you can hear clearly over headphones or monitor speakers how it sounds as you play.

If your engineer doesn't play in the band, it's better to use two rooms: a studio for the musicians and a control room for the recording equipment. In the control room the engineer operates the recording equipment and listens through the monitor system to the sound picked up by the microphones. If you can record this way, run a microphone snake from the studio to the control room. This snake will carry the signals from the mic cables back to the mixer.

Acoustics

Ideally, the studio should be large and quiet and not square. A square room responds strongly to certain low notes, making them "boom out." You may have to use whatever room is available; that's okay, but your recording results may be less than optimum. To keep out noise, put weatherstripping under doors, close doors and windows, and turn off air conditioning while recording.

Unless you're recording classical music, the room should be acoustically dead—free of echoes and reverberation. To further deaden the room, carpet the floor and add some absorbers (described in Chapter 2).

Equipment Layout for a Multitrack System

One-Room System

Figure 3-1 shows how to lay out a one-room recording system. On a large table, place your mixer and multitrack recorder (or recorder-mixer), the two-track recorder (open-reel or cassette), signal processors, monitor speakers, and headphones. You might want to put this equipment on a large sheet of plywood for storage under a bed or in

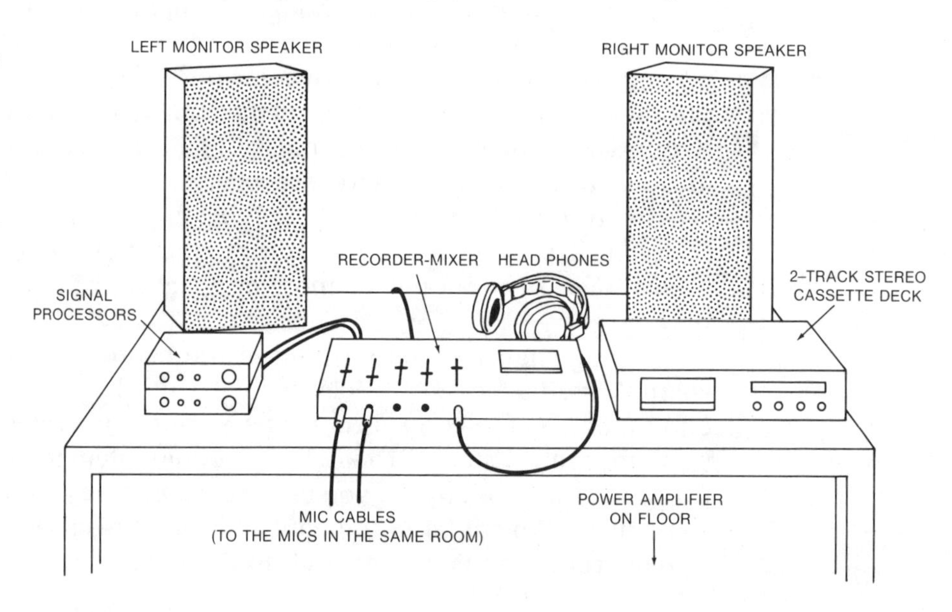

Figure 3-1. Typical layout of a one-room recording system.

a closet. Place the stereo power amplifier for driving the loudspeakers anywhere that it can be air-cooled, but close enough to your loudspeakers to enable you to run short cables between amp and speakers. Long cables waste power.

If you're playing a piano or synthesizer ("synth"), you might be able to place a recorder-mixer on top of your instrument, on your keyboard stand, or on a table nearby.

Monitor speakers should be high-fidelity bookshelf types costing at least $100 each or small studio monitor loudspeakers. To reduce the influence of room acoustics on the sound of the speakers, place the speakers about 3′ apart and 3′ from you—an arrangement called *close-field monitoring* (shown in Figure 3-2). Sit exactly between them to perceive correct stereo imaging.

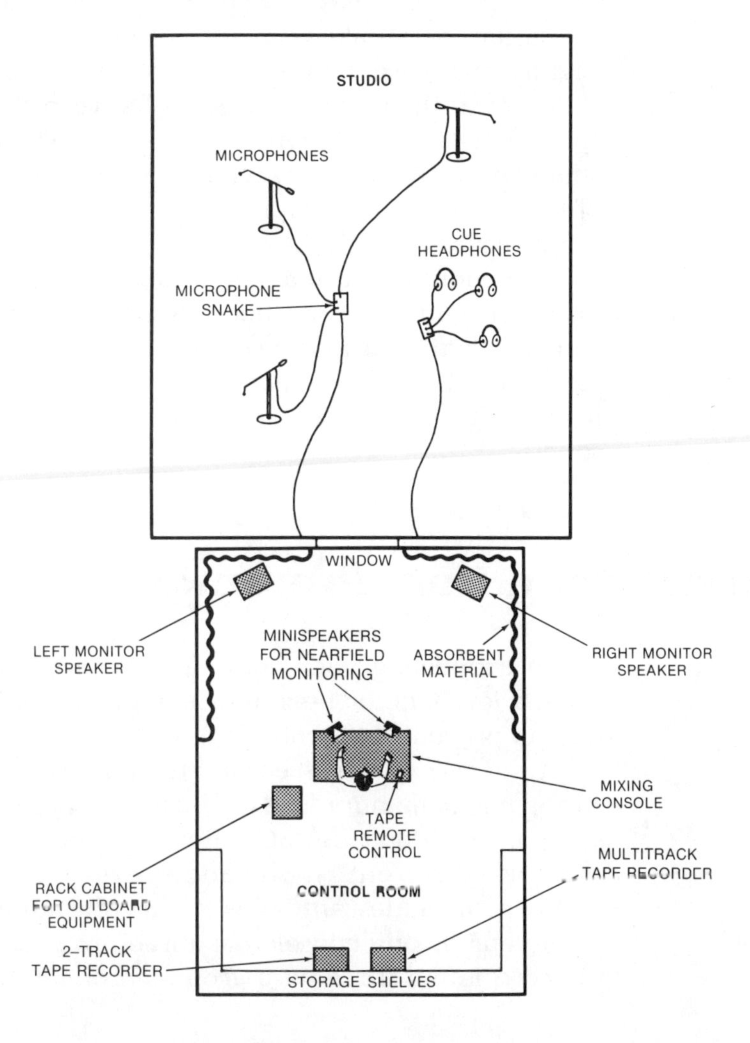

Figure 3-2. Typical layout of a two-room recording studio.

Small speakers can be mounted in several ways:

- on the table top
- on top of the mixer over the meters (if your mixer is large)
- on keyboard stands

Two-Room System

For a more elaborate setup, with studio and control room, you could locate the recorders at the rear of the control room with remote controls at the mixer position (as shown in Figure 3-2). Or just set up your control room as suggested for the one-room system. Place the signal processors within easy reach, perhaps in a rack to the side.

Mount the speakers on shelves or tables so that the tweeters are slightly above ear height. Place the speakers as far apart as you're sitting from them, and sit exactly between them to perceive correct stereo imaging (as shown in Figure 3-2). The control-room acoustics should be fairly dead. Put acoustical absorbers behind and to the sides of the speakers. This improves stereo imaging and results in clearer, more accurate monitoring.

You could also use a pair of quality headphones for monitoring. In fact, if you hear too much of the live sound of the band in the control room, you may *need* to wear headphones to block out that sound and hear only what you're recording. Headphones also eliminate the problem of control room acoustics coloring the monitored sound.

Powering and Hum Prevention

When connecting your equipment, you run the risk of picking up hum: a low tone or buzz in the audio signal. This section describes how to prevent this problem.

Avoid using fluorescent lights in the studio because they radiate strong magnetic hum fields, which can cause hum in your recording system. Also don't use SCR dimmers (those that use silicon control rectifiers)—they put "hash" and buzzes on the ac line. Instead, use multiway incandescent bulbs to vary the studio lighting levels.

If your mixing console has three prongs on its power cord, plug the cord into a three-wire grounded outlet. Such an outlet has two slits for the power and a U-shaped hole for the power ground. Many homes or buildings with older wiring do not have third-wire power grounds. In that case, try to ground to a stake or pipe driven into the earth. Run a thick insulated wire (#4 is ideal) from the mixer chassis to the stake. Securely bond the ground wire to the stake with a pipe clamp (available at hardware stores).

Plug all components into one or more outlet strips fed from the same circuit breaker or fuse after making sure that the current requirement for the complete system (the sum of the equipment fuse ratings) doesn't exceed the amperage rating for that breaker. These components include both the recording equipment and the instrument amplifiers. Power only your audio equipment from the single breaker; if the breaker also powers a refrigerator or other motorized devices, they may create clicks or noises in your audio signal. Run a long extension cord from the control-room outlet strip to the studio outlet strip.

Test each piece of audio gear having a two-prong power cord to find the minimum-hum position. Proceed as follows:

1. Find an electrical ground such as a metal cold-water pipe, the U-shaped hole in three-hole wall outlets, or the metal screw that holds the cover plate to the wall outlet. Check to ensure the U-shaped hole or metal screw is grounded by connecting a neon tester between the hole or the screw and the outlet sockets. If the tester glows in either of the sockets, the hole or screw is grounded. If not, use the cold-water-pipe ground for the following procedures.
2. Unplug all audio cables and ground leads from the component under test. Turn it on. Connect the neon tester between the component's chassis and the ground (from step 1). If the tester glows, reverse the ac power plug in the outlet. The position causing the least glow is the correct one. Mark the proper polarity on all outlets and equipment plugs.

Audio Cables

Cables carry electrical signals from one component to another. The cables are usually made of one or two insulated conductors (wires) surrounded by a fine-wire shield which reduces hum. Outside the shield is a plastic or rubber insulating jacket.

Cables are either balanced or unbalanced. A *balanced line* is a cable that uses two conductors to carry the signal, surrounded by a shield (as shown in Figure 3-3). An *unbalanced line* (shown in Figure 3-4) has a single conductor surrounded by a shield.

Recording equipment also has balanced or unbalanced connectors. Be sure your cables match your equipment. Balanced equipment has a three-pin (XLR-type) connector (shown in Figure 3-5); unbalanced equipment has a ¼″ phone jack or an RCA phono jack connector (shown in Figures 3-6 and 3-7). A jack is a receptacle; a plug inserts into a jack.

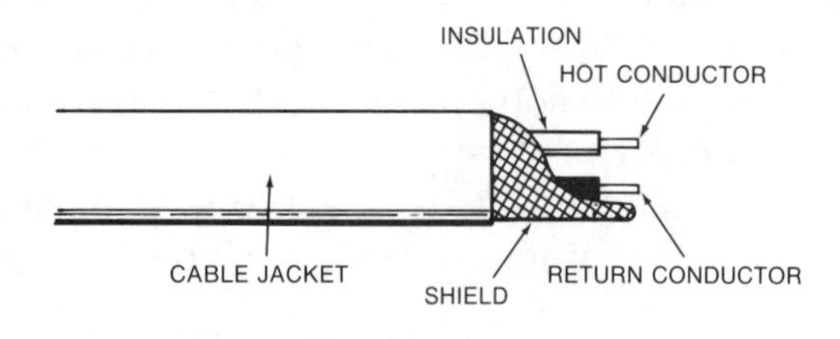

Figure 3-3. Two-conductor shielded, balanced line.

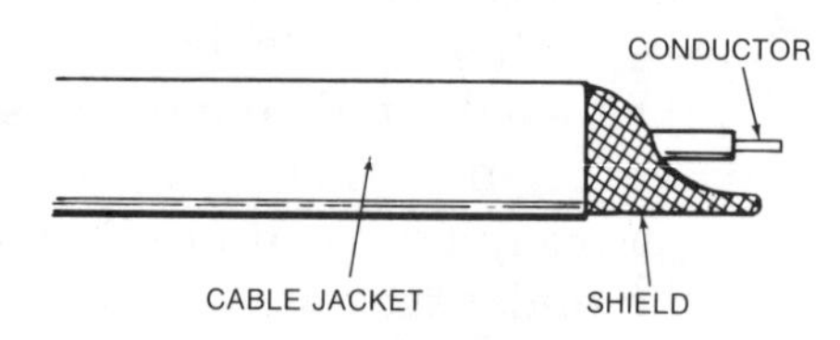

Figure 3-4. One-conductor shielded, unbalanced line.

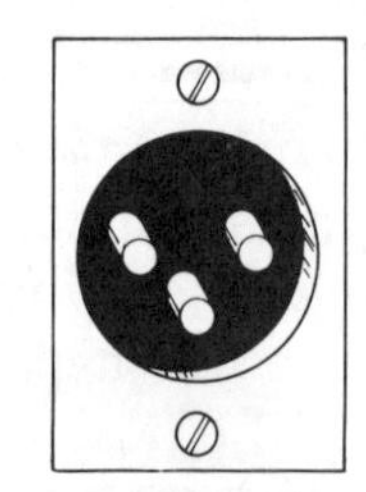

Figure 3-5. Three-pin XLR-type connector used in balanced equipment.

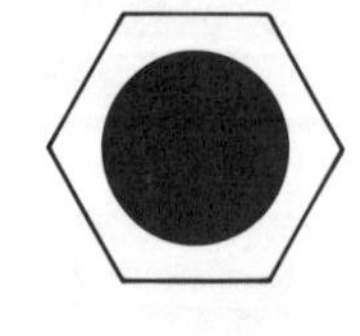

Figure 3-6. A ¼″ phone jack used in unbalanced equipment.

Figure 3-7. RCA phono jack used in unbalanced equipment.

The balanced line rejects hum better than an unbalanced line, but an unbalanced line under 10′ costs less and usually provides adequate hum rejection.

A cable carries one of these three signal levels or voltages:

- Mic level (about 2 millivolts [mV], or 0.002 volt [V])
- Line level (0.316 V or −10 decibels relative to 1-volt [dBV] for unbalanced equipment, 1.23 V or +4 decibels relative to 1-milliwatt [dBm] for balanced equipment). There's a 12-dB difference between these two signal levels.
- Speaker level (about 1 to 1,000 watts [W] or about 20 V).

When recording, keep microphone cables physically separated from line-level cables, and keep these separated from speaker cables and power cords. Noise or hum may occur if high-level and low-level signals are run together.

Speaker cables are normally made of lamp cord (zip cord). To avoid wasting power, speaker cables should be as short as possible, and should be heavy gauge (between 12 and 16 gauge). Number 12 gauge is thicker than 14; 14 is thicker than 16.

Cable Connectors

There are several types of connectors used in audio. Figure 3-8 shows a ¼″ phone plug, which is used with cables for unbalanced microphones, synthesizers, and electric instruments. The tip terminal is soldered to the cable's center conductor and the sleeve terminal is soldered to the cable shield.

Figure 3-9 shows an RCA or phono plug, which is used to connect unbalanced line-level signals. The center pin is soldered to the cable's center conductor and the cup terminal is soldered to the cable shield.

Figure 3-10 shows a three-pin (XLR-type) professional audio connector, which is used with cables for balanced microphones and balanced recording equipment. The female connector (with holes, Figure

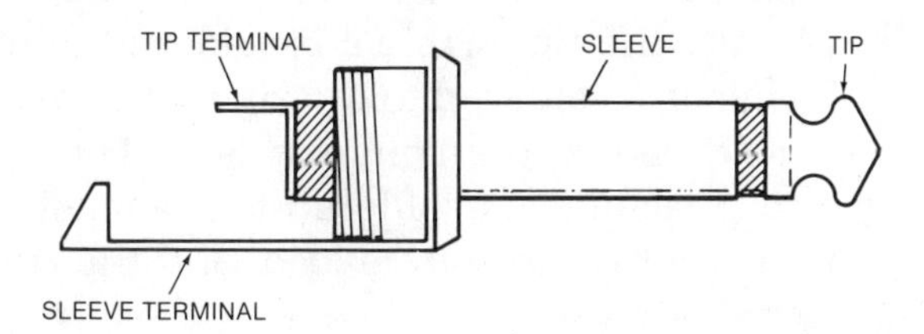

Figure 3-8. ¼″ phone plug.

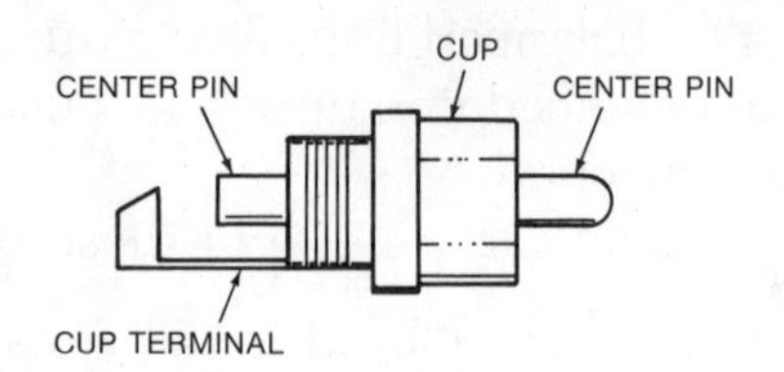

Figure 3-9. RCA phono plug.

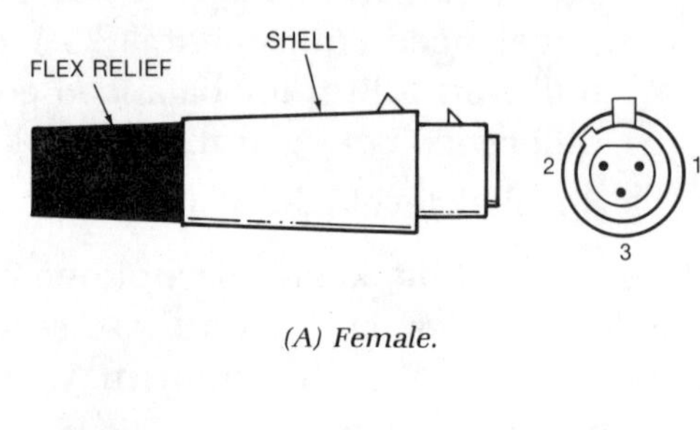

(A) *Female.*

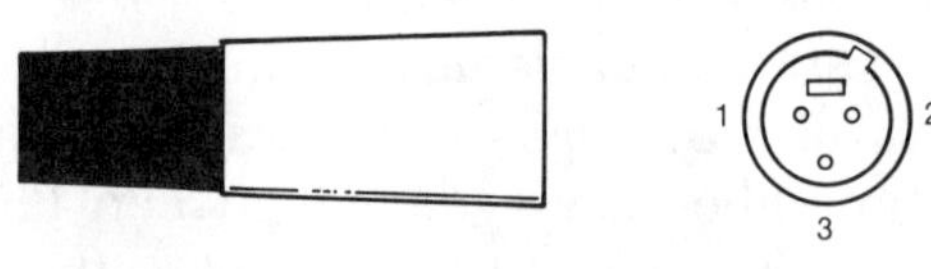

(B) *Male.*

Figure 3-10. XLR-type connectors.

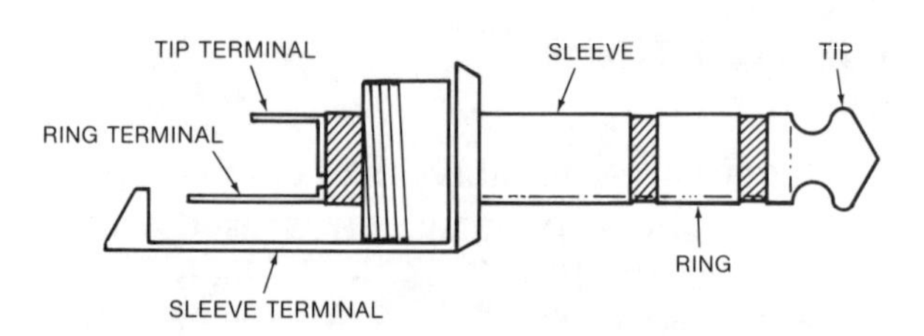

Figure 3-11. Stereo phone plug.

3-10A) plugs into equipment outputs. The male connector (with pins, Figure 3-10B) plugs into equipment inputs. In both female and male connectors, pin 1 is soldered to the cable shield, pin 2 is soldered to the "hot" red or white lead, and pin 3 is soldered to the remaining lead.

Figure 3-11 shows a stereo phone plug, which is used with stereo headphones and with some balanced line-level cables. For headphones, the tip terminal is soldered to the left-channel lead, the ring terminal is soldered to the right-channel lead, and the sleeve terminal is soldered to the common lead. For balanced line-level cables, the sleeve terminal is soldered to the shield, the tip terminal is soldered to the hot red or white lead, and the ring terminal is soldered to the remaining lead.

If you have unbalanced microphone inputs on your recorder or mixer, use a balanced cable from mic to input to reduce hum. Solder the shield and black lead to the long ground lug (sleeve terminal) on the phone plug. Solder the white or red lead to the small center lug (tip terminal) on the phone plug (as shown in Figure 3-12).

When you connect balanced equipment to unbalanced equipment, use the connection shown in Figure 3-13 to compensate for level differences. Balanced equipment operates at a relatively high level or voltage (called " +4 dBm" or " +4"); unbalanced equipment operates at a lower level (called "−10 dBV" or "−10").

Reducing Microphone Hum

Microphones and mic cables are especially sensitive to hum pickup because of the great amplification needed for mic-level signals. Here are some tips to minimize microphone hum pickup:

- Use low-impedance microphones (150–600 ohms [Ω]), which pick up less hum than high-impedance microphones. All modern recording equipment is designed to work with low-impedance microphones, whether balanced or unbalanced. If you plug a high-impedance microphone into a low-impedance

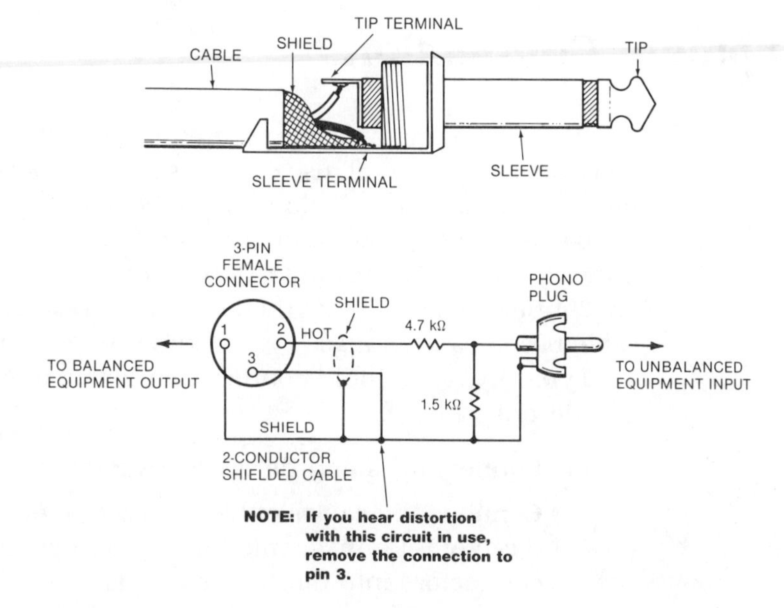

Figure 3-12. Wiring a balanced mic cable to an unbalanced ¼″ phone plug.

Figure 3-13. A 12-dB pad for matching a balanced +4-dBm output to an unbalanced −10-dBV input.

input, the signal will be weak, the bass will probably sound thin, and the microphone may distort.

- Use microphones with balanced outputs (three-pin connectors), which pick up less hum than unbalanced microphones (which use a one-conductor shielded cable). If your mic cable is under 25′ long, however, an unbalanced mic will probably be adequate.
- Routinely check mic cables to make sure that the shield is connected at both ends.
- In each microphone, look for a set screw in the handle near the XLR-type connector. Check that the set screw is securely screwed clockwise (out) into the mic handle.
- Inside the XLR-type connector is a ground lug that makes contact with the connector shell. If the cable is used in a studio, solder the ground lug to pin 1 so that the shell is grounded and acts as a shield. If the cable is used in remote or outdoor setups, do not connect the ground lug to pin 1 because hum may occur if the shell touches a metallic surface.
- Do not ground microphone junction boxes or snake boxes except through the cable shield.

Equipment Connections

The instruction manuals for your equipment tell how to connect each component. In general, use cables as short as possible to reduce hum, but long enough to be able to make changes.

Be sure to label all your cables on both ends according to what they plug into. For example, "Mixer ch. 1 monitor out," or "Reverb ch. 2 input." This way, if you temporarily change connections, or the cable becomes unplugged, you'll know where to replug it.

Typically, you would connect equipment as shown in Figure 3-14 and described here.

1. Connect mics and direct boxes to mic cables.
2. Connect mic cables either to the snake junction box or directly into mixer mic inputs. Connect the snake connectors into mixer mic inputs.

Figure 3-14. Equipment connections.

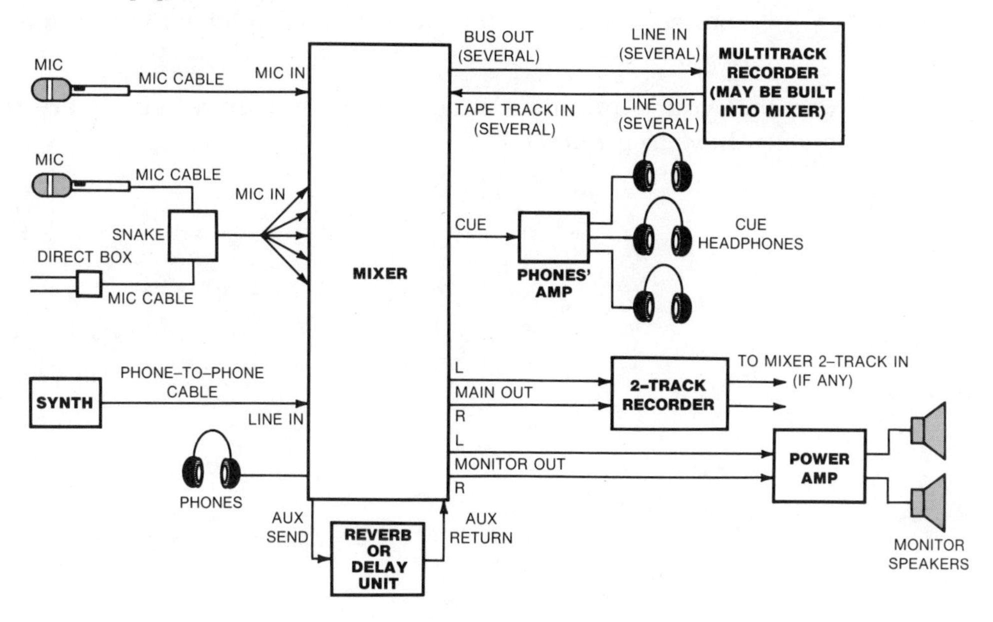

3. Connect synthesizers and drum machines to mixer line inputs. If this causes hum, use the direct-connection cable shown in Figure 2-6.
4. Connect mixer main outputs to a 2-track recorder.
5. Connect 2-track recorder outputs to mixer 2-track inputs (if any).
6. Connect mixer monitor outputs to power-amplifier inputs.
7. Connect power amplifier outputs to loudspeakers.
8. Connect mixer aux-send connectors to digital delay or reverb inputs.
9. Connect delay or reverb outputs to mixer aux-return or bus-in connectors.
10. If you're using a separate mixer and multitrack recorder, connect mixer bus 1 to recorder track 1 in, connect bus 2 to track 2, and so on. Also connect recorder track 1 out to mixer tape-track 1 in, connect track 2 out to mixer tape-track 2 in, and so on.

11. If the musicians in the studio are using headphones, connect the cue output to a small amplifier to drive their headphones.

Now that your equipment is connected and hum-free, you're ready to learn how to use it.

4 Recording a Soloist or Small Acoustic Group

A band playing rock, jazz, or pop music may require multiple microphones and a mixer to achieve a commercial sound. But some musical groups can be recorded effectively with just a stereo cassette deck and two microphones. The simpler methods work well with quartets, soloists, ethnic bands, folk groups, orchestras, pipe organs, or symphonic bands. With a little practice, even beginners can make quality recordings with two mics and a cassette recorder.

The recording should be as clear, realistic, and as noise-free as possible. In addition, the balance among instruments and voices (their relative loudness) should be appropriate for the music performed. This chapter offers some tips to achieve these goals.

Equipment

Good recordings can be made with simple equipment.

- A quality stereo cassette deck such as that used in a home stereo system.
- Blank cassette tape—brand-name; high-bias, CrO_2, or metal; C-60 length or shorter. Use the tape recommended by the cassette-deck manufacturer.
- A stereo microphone costing at least $50, or two separate high-quality microphones.
- One or two mic stands and booms.

Cassette Deck

Let's consider the requirements for a quality cassette deck. Obtain the published specifications for the deck you want to buy or use, and check its adherence to the following capabilities.

Noise Reduction

This is a circuit that reduces tape hiss. The Dolby and dbx systems are common. Noise reduction is essential with cassette recorders because the slow tape speed and narrow track width result in audible tape noise. Noise reduction circuits clean up the signal.

Dolby C is more effective than Dolby B and dbx is more effective than either. Still, Dolby is free of the "breathing" (modulation) noise that is sometimes heard on dbx'd tracks. Chances are you'll be equally satisfied with either Dolby C or dbx.

Wow & Flutter

Wow is a slow periodic variation in tape speed; *flutter* is a rapid variation. If excessive, each causes the pitch of recorded instruments to wobble. The lower the wow-and-flutter spec, the steadier the reproduced pitch:

- 0.03% RMS weighted (or WRMS) is excellent.
- 0.04% RMS weighted (or WRMS) is very good.
- 0.1% IEC/ANSI peak weighted is very good.
- Higher values are not as good. They indicate that you might hear the pitch wobble on recordings of fretted stringed instruments or piano.

Signal-To-Noise Ratio

This is the ratio, expressed in dB, between the maximum undistorted recorded signal level and the noise level. The higher this figure is, the more noise-free the recording is. All the following specs are measured with noise reduction:

- 90 dB is excellent (typical of dbx).
- 70 dB is very good (typical of Dolby C).
- 65 dB is good.
- 55 dB is fair.

These specs are A-weighted, which means that the measurement was done in a way to correlate with the annoyance value of the noise.

When you are comparing two decks, be sure that both signal-to-noise (S/N) specs are A-weighted.

Record-Play Response

This is the range of frequencies that the recorder will record and play back at an equal level, within a tolerance (such as ±3 dB). The lower the lower frequency, and the higher the upper frequency, the better the fidelity.

- 40 Hz–18 kHz ±3 dB is excellent.
- 40 Hz–14 kHz ±3 dB is good.
- 40 Hz–12.5 kHz ±3 dB is fair.

Microphones

Now let's consider microphones. Use a stereo microphone costing at least $50 if you're recording a singer-guitarist; a classical-music soloist; or a small acoustic group, such as a vocal quartet or folk group. An alternative to a stereo mic is two identical microphones. If these are the cardioid type (preferred), mount them on a *stereo bar*—a device that holds two mics on a single stand for stereo miking. Angle them apart about 110 degrees (°) or about 55° either side of center, and space their grilles 7″ apart horizontally, as shown in Figure 4-1.

If the two identical microphones are the omnidirectional type, place each on a mike stand and space them 3′ apart for a small group or 10′ apart for a large symphonic ensemble.

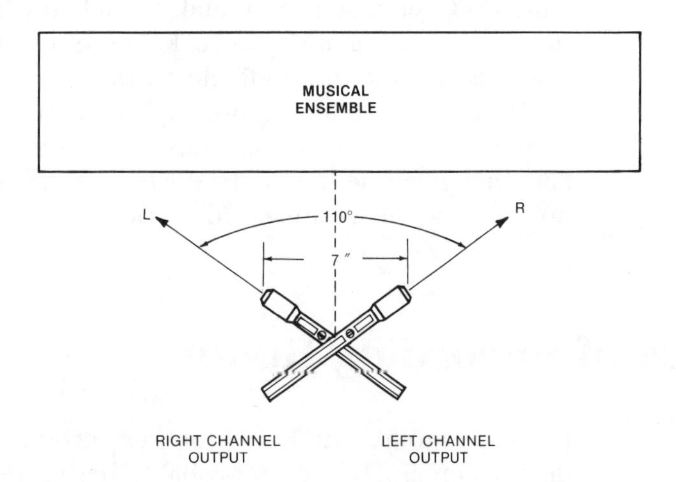

Figure 4-1.
A stereo miking method.

For a singer who accompanies him- or herself on piano, use an omnidirectional microphone for the voice. Unlike a directional microphone, an omnidirectional unit does not get bassy when you place it close to the mouth. Chances are that an electret condenser type will sound best. It requires use of a battery, which lasts up to a year. If you already have another type of microphone, it's okay to use it, but an omnidirectional condenser mic is recommended. For the piano, use any mic you have if it sounds good to you, or one of the following:

- a miniature omnidirectional condenser microphone
- a cardioid electret-condenser microphone
- a PZM surface-mounted microphone

Prerecording Setup

Inside the cassette door are *heads,* metal blocks that contact the cassette tape. They either erase or record and play back sounds on tape. If the heads get dirty, the sound becomes dull. So before each recording clean the heads with a cotton swab moistened with the cleaning fluid recommended in your recorder manual. Or use isopropyl or denatured alcohol (from a drugstore). *Do not* use rubbing alcohol. Gently rub the swab on the head surfaces that contact the tape, and then rub them dry. Clean the rubber roller, too.

Set the tape-type switch to the type of tape that you're using (it's specified on the cassette). As we noted earlier, this should be high-bias, CrO_2 (chrome), or metal. Switch on the noise reduction, both during recording and playback. Some decks let you choose between Dolby B and the more effective Dolby C.

If you plan to send your tape to someone else, find out whether their cassette machine has noise reduction and, if so, what type it is. Set your machine to match. Write on the cassette label which kind of noise reduction you used (dbx, Dolby B, or Dolby C).

Choice of Recording Room

If you're taping a folk or bluegrass group, record in an acoustically dead room. Such a room probably has carpeting, acoustic-tile ceiling, stuffed furniture, and drapes. If you need to deaden the room further, add absorbers like those described in Chapter 2.

If you're recording a classical-music soloist or ensemble, record in a "live" room that has noticeable reverberation, such as a church or recital hall. The room acoustics enhance the recording for this type of music.

Microphone Techniques

Screw the microphone stand adapter onto a mic stand. Place the mic in its stand adapter.

When miking a soloist or ensemble in stereo, put a stereo microphone (or a pair of identical microphones) close to the musicians, from 2–5′ away, as in Figure 4-2. Place the mic about 1–2′ away to pick up a singer playing an acoustic guitar. For a grand-piano solo, raise the lid on the long stick.

If you're recording a small acoustic group, you might try the inconspicuous arrangement shown in Figure 4-3. Mount two miniature omni condenser mics 14″ apart on a 2′-square panel (boundary) made of plexiglass or ¼″ plywood. The front of each mic aims at the panel and is spaced 1/32″ from the panel. Place this array on the floor, angled up at the performers about 5′ away (as in Figure 4-4), in front of the center of the group. If only two people are playing in a song, put the array in front of and between those two people.

To record a singer who accompanies him- or herself on piano, you'll need two mics—one for the voice, one for the piano. You'll also need one or two booms. A boom is an adjustable pipe that mounts on

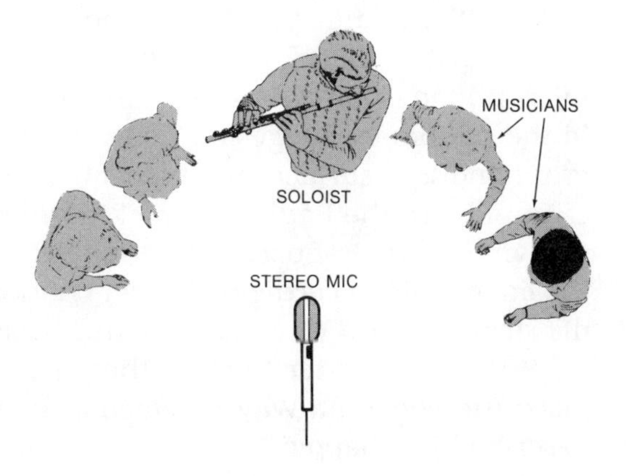

Figure 4-2. Miking a soloist or ensemble with a stereo microphone (top view).

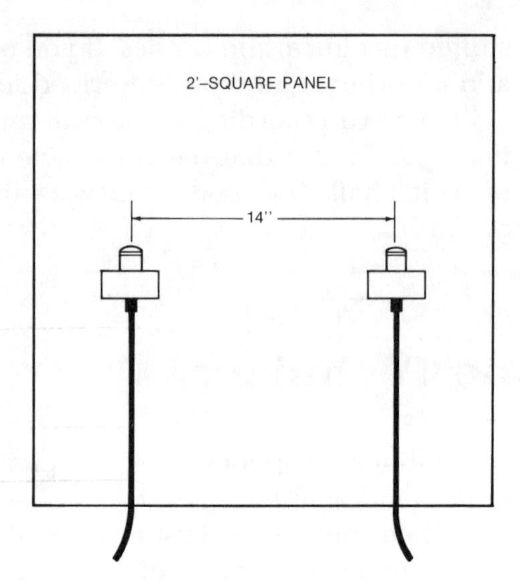

Figure 4-3. A stereo miking array using a panel and two mini omni condenser mics.

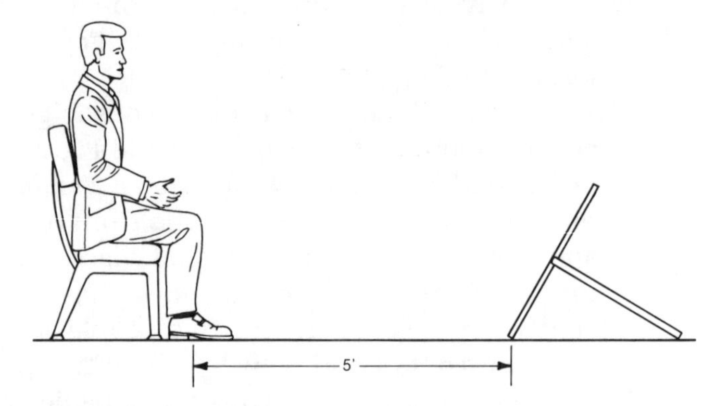

Figure 4-4. Boundary-mic array placement.

a mic stand for positioning the mic. Use a foam pop filter or windscreen on the vocal microphone to reduce breath "pops." Place the mic about 1″ from the mouth (as is shown in Figure 4-5).

For grand piano, you might want to use a miniature condenser microphone (such as is shown in Figure 4-6). Tape it to the raised lid in the middle, as suggested by the manufacturer. Another useful microphone is a surface mounted microphone—a flat-plate unit. Some examples are the Radio Shack PZM units selling for $15 to $45, or the Crown Sound Grabber, selling for $49 to $99. Alternatively, remove the lid and aim a cardioid electret-condenser mic down over the middle strings, about 1′ up, and about 1′ horizontally from the hammers.

You might prefer to mike the piano in stereo and use a mixer to place the vocal midway between your stereo speakers, a technique described in Chapter 7.

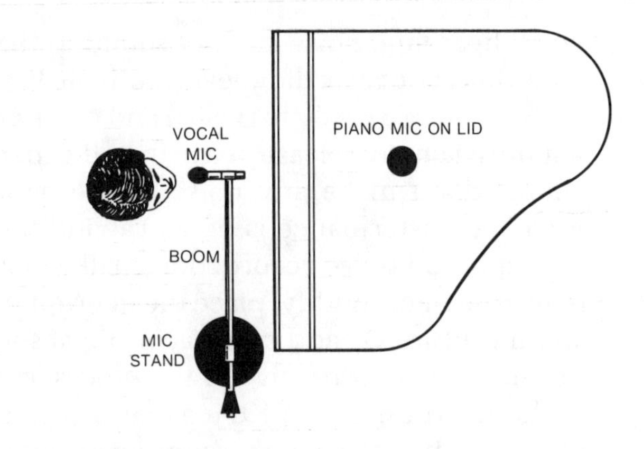

Figure 4-5.
One way to mike a singer and piano (top view).

Figure 4-6.
Crown GLM-100 miniature condenser microphone.
(Courtesy of Crown International)

Recording

Plug the mic cables into the left and right mic inputs of your cassette deck. Press the record and pause buttons. Follow these steps to set a proper recording level. The recording level is indicated by a meter or column of lights that show how loud the sound is. While recording music, gradually turn up the recording-level control (like a volume control) until the meter reads around 0 (just below the red area) on the loudest sounds. Occasional excursions into the red might be okay, depending on the tape used.

If the meter reading is too high (frequently into the red), you'll hear distorted sound on playback. If the meter reading is too low, you'll hear tape noise (a hissy sound in the background). Experiment with different recording levels to hear these effects.

Now you're ready to record and make adjustments. Press the pause button again to release it so that the deck starts recording. While taping, don't make any noise before or after performing the song. After the performance is done, rewind the cassette and listen to it.

Suppose you've recorded a small acoustic ensemble. If the sound is too distant or muddy, place the microphone(s) closer to the ensemble and try again. Or add more acoustic absorbers to the room. If a musician is too quiet relative to the others, have him or her move closer to the microphone and try again. If the balance still is poor or the recording has too much room acoustics, try miking voices and instruments up close and blending them with your mixer. This technique is described in Chapter 7.

Now suppose you've recorded a singer who plays guitar. Note the suggestions above about miking distance. Raise the mic on its stand if the voice is too quiet; lower it if the guitar is too quiet.

Let's say you've recorded a singer who plays piano. If the voice is too loud relative to the piano, turn down the volume control slightly for the vocal microphone and try again. Or turn down the piano mic if the piano is drowning out the vocal.

When you're satisfied with the balance and microphone distance, record other tunes. Leave about 4 seconds of silence between songs. That's your finished master tape!

Tape Copies

If you want to make copies of your master tape, you can either copy from one deck to another, or you can use a dubbing cassette deck that holds two cassettes.

To copy from one deck to another, connect a cable between the output connectors of the playback deck and the record-deck input connectors of the recording deck. This cable should have RCA phono plugs on each end to match your equipment. Set the recording level carefully so that it will peak around 0 on the loudest parts.

To copy a tape with a dubbing deck, insert your master tape into the "play" section and insert a blank chrome or metal tape into the "record" section. The dubbing deck might work at two speeds; the slower speed usually provides better fidelity. If necessary, set the recording level and press the record button to copy the tape.

Whether using a dubbing deck or two decks for your copy, be sure to set the tape-type switches and noise-reduction switches appropriately. Ideally, the copy will sound nearly as good as the master tape.

5 Recorder-Mixer Features

If you want to record pop music, rock, or jazz, you'll need to learn the technology of mixers and multitrack recorders. This equipment is required to produce a professional, commercial sound. All the knobs, switches, and lights can be intimidating at first, but you'll understand them easily enough if you read the instruction manuals and practice with the equipment.

Why is multitrack recording equipment so complicated? Why can't you just plug in a microphone and hit the record button? The reason is that there are many characteristics you want to control.

- the relative loudness of each instrument (the balance)
- the tone quality of each instrument (bass, treble, midrange)
- the space that the instruments are in (reverberation)
- the left-to-right position of each instrument (panning)
- special effects (flanging, echo, chorus, etc.)

While you're recording, you also need to control which instrument goes on which track. If you have more instruments than tracks, you need a way to mix or combine several instruments onto one track. In addition, you must adjust the recording level or volume so that the recorded signal is not distorted or noisy.

You also need to control what you're listening to or monitoring.

- During recording, if you're mixing several instruments to one track, you need to hear all the instruments mixed (as in the final product).
- During playback, you want to hear the tape tracks mixed to approximate the final product.
- During overdubs, you want to hear prerecorded tracks mixed with the live instrument to be overdubbed.

- During mixdown, you want to hear all tracks mixed to two-channel stereo.

With all these control requirements, it's easy to see why modern recording equipment is complicated and requires some study to understand its functions. Fortunately, equipment for home studios is much simpler than equipment for professional studios.

Overview of Recording, Overdubbing, and Mixdown

Before describing recorder-mixer features, we need to know the three stages in making a multitrack recording.

1. Recording—you record one or more instruments onto one or more tracks. Usually the rhythm instruments—drums, bass, guitar—are recorded first.
2. Overdubbing—while listening to prerecorded tracks over headphones, musicians add new parts that are recorded on open (unused) tracks. Vocals and quiet acoustic instruments are usually overdubbed.
3. Mixdown—once all the tracks are recorded, you mix or combine them into 2-track stereo. This stereo mix is recorded onto an external 2-track cassette or open-reel deck. This is the final master tape, which can be duplicated on cassettes or records.

Mixer Section of the Recorder-Mixer

Let's examine the features of the recorder-mixer in detail.

The mixer portion has several functions. It

1. amplifies the signals from all the microphones
2. mixes them in various combinations
3. routes them to tape tracks
4. adjusts the tone quality and stereo position of the instruments
5. controls special sound effects

The mixer can be divided into three sections: input modules, output module, and monitor section.

Features of the Input Module

An input module (See Figure 5-1) is a group of controls in the mixer that affects a single input signal. Each module is usually a narrow vertical strip. In cassette recorder-mixers, the number of input modules ranges from two to eight, with four modules the most common number. The more input modules available, the more mics you can use during a recording. If you're recording yourself only, you may need only two inputs.

Figure 5-2 shows the signal flow from input to output through a typical input module and through the other circuits in a mixer. Each component of the diagram is described here, generally following input to output, or from left to right.

Input Connectors

These connectors are for microphones and other signal sources. In some recorder-mixers, a single ¼″ phone jack is used both for mic-

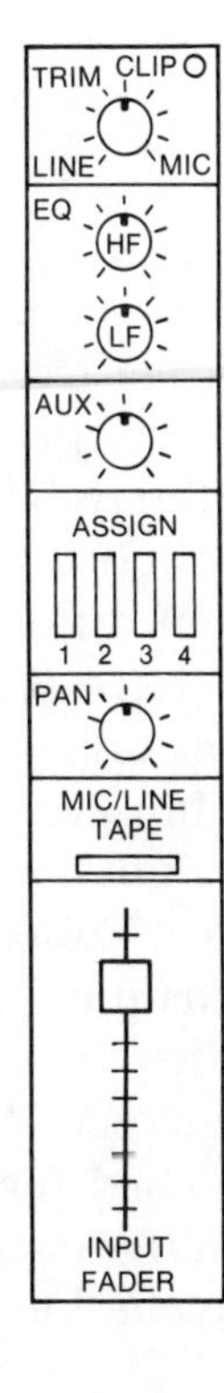

Figure 5-1. A typical input module.

Figure 5-2. Signal flow in a typical mixer section of a recorder-mixer.

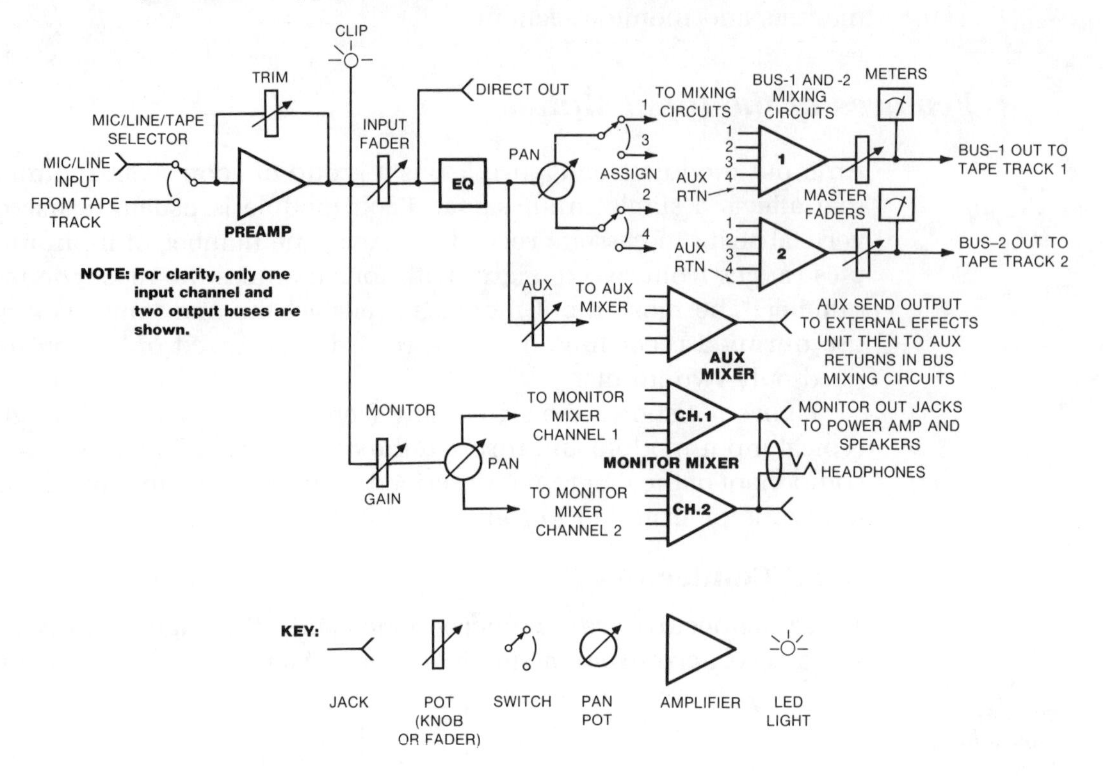

level and line-level signals. A mic-level signal (from a microphone) is about 1–2 mV nominal. A line-level signal (from a synthesizer, drum machine, or direct from an electric guitar) is about 0.3–1.23 V.

If a single jack is provided for both mic and line signals (as is shown in Figure 5-2), the two levels are handled either by a mic/line switch; a hi-lo gain switch; or a trim control, which reduces the line-level signal to prevent distortion.

Some units have separate jacks for mic and line inputs. The mic inputs are either unbalanced ¼″ phone jacks or XLR-types. The phone jacks cost less but are more susceptible to pickup of buzz and hum, although this is not a serious problem in most small studios. The line jack is either a ¼″ phone jack or an RCA phono jack. You can plug an electric instrument directly into such an input without using a direct box, if the cable is short enough to prevent hum.

Some newer recorder-mixers have a sync input, which goes into track 4 and is used for recording a special tape-sync signal from a computer running a sequencer program. It's becoming common, in the final mix, to combine sequencer recordings of MIDI-equipped synthesizers with tape recordings of real instruments and vocals. The tape-sync feature lets you synchronize tracks recorded on tape with sequencer tracks recorded in computer memory. It also lets you overdub two or more sequencer tracks onto tape by keeping them synchronized. More about this in Chapter 12.

Input Selector

This switch selects the input you want to process: mic, line, or tape. The switch usually is labeled in one of these ways:

- mic/line/tape
- mic/line/remix
- mic-line/off/tape
- input/mute/track
- input/off/line

The various switch positions work as follows:

- Mic—the mic signal enters the mixer.
- Line—the line signal enters the mixer.
- Mic-line or Input—either the mic signal or the line signal enters the mixer, depending on what is plugged into that input.
- Tape, Track, or Remix—the tape-track signal enters the mixer (for overdubbing or mixdown).
- Off or Mute—no signal is processed. To reduce tape hiss during mixdown, it's a good idea to mute tracks that have nothing playing at the moment.

Microphone Preamplifier (Preamp)

This amplifier inside the input module boosts the weak microphone signal to a stronger line-level signal.

Trim (Gain)

Trim is an extra volume control that adjusts for a wide range of input signal levels. You can set the trim control in two ways: by observing the LED (light-emitting diode) overload indicators or by observing the meters. *LED overload* (*clip* or *peak*) indicators are little lights that flash when the input signal level is so high that it is causing distortion in the mic preamp. This occurs when a microphone is picking up a very loud instrument. If the overload light flashes, gradually turn down the trim control until the light stays off.

Another way to set the trim control is to set the master fader(s) to the shaded portion of its range (sometimes at 0, about ¾ up). Do the same for the input fader. Assign the signal to a tape track, and adjust the trim control so that the meter peaks around 0.

Inexpensive recorder-mixers do not have overload LEDs or a trim control.

Input Fader

After the microphone signal is amplified by the preamp, it goes to the *input fader,* a sliding volume control for each input signal. Generally, you use it during recording to set recording levels on tape and during mixdown to set the relative loudness balance among instruments. (In the Fostex X-15 II, the input faders are used during recording to set recording levels and during mixdown as master volume controls.)

Direct Out

Only the more elaborate units have this feature: an output connector that follows each input fader. The fader controls the level at the direct-output jack. Direct out is used for recording one mic on each track of an external multitrack recorder. Since direct out bypasses the mixing circuits farther on in the chain, the result is a cleaner signal.

Equalization (EQ)

The signal from the input fader goes to an equalizer, which means "tone control." With EQ, by boosting or cutting certain frequencies, you can give an instrument's sound more or less bass or more or less treble.

Equalizers cover two or three bands of the audible spectrum. Inexpensive units have simple two-knob bass and treble controls; you can boost or cut the treble or bass. Fancier models have a sweepable EQ (sometimes called "parametric" EQ) that lets you "tune in" the exact frequency range you want to work on. This feature adds cost and complexity, but it gives you more control over the tone quality.

In inexpensive units, the EQ works on two tracks at a time during recording, and on the stereo mix during mixdown. This is less flexible than a unit with EQ on each input.

Assign Switches

The equalized signal goes to the *assign* switches or knobs. These let you assign the signal of each instrument to the tape track you want to record that instrument on. Some units have a *track selector* switch labeled "1, 2, 3, 4." Others assign tracks by using a combination of the pan pot setting (explained next) and the record-select switches. Some inexpensive units always assign input 1 to track 1, input 2 to track 2, and so on.

Pan Pot

During recording and overdubbing, the *pan pot* is often used to assign inputs to tracks: left for odd tracks, right for even tracks. For example, if you set the pan pot left, you can record on tracks 1 or 3, or both, depending on which record-select button you press. During mixdown, the pan pot places the stereo image of each recorded track wherever desired between a pair of stereo loudspeakers. Consequently, you can locate an instrument at the left speaker, right speaker, or anywhere in between.

Aux (Effects or FX)

The *aux* control is shown in Figure 5-2 just after the EQ. You use this knob during mixdown to control the level of an input signal sent to an external effects device, such as a digital delay or reverberation unit. The processed signal returns to the mixer, where it blends with the original signal, adding a sense of ambience or spaciousness to an otherwise "dry" track. This feature is essential if you want to produce a commercial sound.

During recording and overdubbing, you can also use the aux knobs to create a mix heard over headphones. In that case, you don't connect the *aux-send* output to an effects unit, but, instead, to a small amplifier that drives headphones. The headphone mix done with the aux knobs is independent of the levels going on tape.

Some units have no aux feature; some have one aux-send control; some have two or more. The more aux sends you have, the more you can play with effects, but the greater the cost and complexity.

A few units have an *aux-receive,* or *aux-return,* control for setting the level of the signal returning to the mixer. The processed effects signal enters the mixer through either a pair of *aux-return* jacks or *bus-in* jacks.

There may be a *pre/post* switch next to the aux-send knob. Use the pre setting for a headphone mix during recording or overdubbing and the post setting for effects during mixdown.

Some high-end recorder-mixers have pan as well as gain for the aux send.

The following input module features are omitted from Figure 5-2 for clarity.

Output Fader (Potentiometer)

In some inexpensive recorder-mixers, this control is found in each input module. It is used as a volume control for that input or track in the monitor or headphone mix.

Access Jacks (Insert Jacks)

These let you plug a compressor in series with an input module's signal for automatic volume control. Inexpensive units omit this feature while some units have access jacks on only two inputs.

The access jacks also can be used to insert a digital delay, or reverb unit, into the signal path of one track. On the delay-reverb unit, you set the dry-delay mix with its built-in mix control. Normally, though, delay-reverb units are patched between the aux-send and -receive jacks, with the mix control on the delay-reverb unit set all the way to "delay," or "wet."

Bounce (Ping-Pong or Transfer)

The *bounce* feature allows you to mix several recorded tracks with your mixer and then record the result on an empty track. Then you erase the original tracks, freeing them for recording more instruments. For example, let's say you've recorded instruments on tracks 1, 2, and 3. You combine these tracks with your mixer and record the mix on track 4. Then you can record three more parts on tracks 1, 2, and 3.

Record/Play/Send Switch

Found in some inexpensive units, this switch works in the following way.

- Record—record this input signal on the same numbered track.
- Play—play this track (or make it safe—not able to be recorded or erased).
- Send—send this track to all the other input modules for bouncing. Only the track with "Record" pressed will record the bounced tracks. For example, if you want to bounce tracks 1, 2, and 3 to track 4, press "Send" for tracks 1, 2, and 3; press "Record" for track 4.

Features of the Output Module

The output module includes mixing circuits, master faders, and meters. In the mixer, the output module is the final section that feeds signals to tape tracks. Let's look at each part in detail.

Mixing Circuits

Each channel or bus that feeds a tape track has one of these circuits. The bus-1 mixing circuit accepts the signals from all the inputs assigned to bus 1 and mixes them in order to feed track 1 of the tape recorder. The bus-2 mixing circuit mixes all the bus-2 assignments, and so on. Mixing circuits also accept aux-return signals, such as the reverberated signal from an external reverb unit.

A four-bus mixer provides four independent output channels or buses; each bus carries a signal that could contain the sounds from one or more musical instruments. The four buses feed a 4-track cassette recorder. By recording two tracks at a time, you can use a mixer with only two output buses with a 4-track recorder.

Master Faders

The output module also contains the *master,* or *output,* faders. These are one, two, or four faders that control the overall level of the output channels.

In a mixing console, the faders that control the level of each bus independently are called *submaster, bus master,* or *group faders.*

Meters

Meters measure the strength (level or voltage) of various signals. Usually, each output bus has a meter to measure its signal level. Since these buses feed the tape tracks, you use the meters to set the recording level for each track. Your recorder-mixer will have one of three types of meters.

- A VU (volume-unit) meter—a voltmeter that shows approximately the relative loudness of various audio signals
- An LED bargraph level indicator—a column of lights (LEDs) that show peak recording level
- An LED peak indicator—a light mounted in a VU meter that flashes when peak recording levels are excessive

The VU meter does not respond fast enough to musical attacks or peaks to indicate them accurately. The LED peak indicator shows true recording levels more accurately.

Tape-Out Jacks

Some units have *tape-out* jacks (not shown). These are connected to the tape-track outputs, and they are used for copying 4-track cassette recordings onto a multitrack studio recorder for further overdubs and processing.

Features of the Monitor Section

The monitor section controls what you hear. It lets you select what you want to hear, and it lets you create a mix over headphones or speakers to approximate the final product. This monitor mix has no effect on the levels going on tape.

Monitor Mixer

The *monitor mixer* (sometimes labeled "monmix" or "tape cue") is a submixer built into the larger mixer. It controls the balance (*monitor mix*) among vocals and instruments heard over headphones or loudspeakers as you're recording.

The monitor mixer is made of several *monitor-gain* and *monitor-pan* controls, plus two mixing circuits that feed either headphones or an external stereo amplifier and speakers.

As shown in Figure 5-2, the monitor gain (volume) and pan for each input signal come before the input fader (prefader) in the signal path. Use the monitor gain knob to control how loud each live instrument or track is in the monitor mix. Use the monitor pan knob to control the position of the monitored instrument or track between your stereo speakers. Some inexpensive recorder-mixers use the output controls on the input modules to create the monitor mix.

The monitor mixer also blends prerecorded tape tracks and live microphone signals into a *cue mix* that is sent to the musicians' headphones in the studio. The musicians record new parts while listening to the cue mix over headphones.

Note that the aux sends in the mixer can serve double duty as controls for a monitor or headphone mix. In most recorder-mixers, the monitor mix and cue mix are identical.

Monitor Select

The *monitor-select* buttons let you choose which signal you want to monitor. Since the configuration of these buttons varies widely among different models of recorder-mixers, they are not shown in Figure 5-2. Here's a listing of some of the monitor select buttons:

- Monitor Tr. 1, 2, 3, 4—you select which track or combination of tracks you want to hear, and mix them with the output controls in the input modules.
- Tape/bus 1, 2, 3, 4—you select whether you want to hear signals off the tape, or from the bus (the mixer output), for channels 1, 2, 3, or 4. Select "tape" to hear a playback or to hear previously recorded tape tracks during an overdub. Select "bus" to hear the live signal that you're recording.
- Tape-bus/stereo/aux or Tape-bus/2-tr/aux—the "tape-bus" switch position is described above. The "stereo" or "2-tr" switch position lets you hear the two-channel stereo mix during mixdown. The "aux" position is used to hear the aux signal (effects or headphone mix).
- Remix/cue/aux—"remix" is the two-channel stereo mix you want to hear during mixdown. "Cue" is the headphone mix. "Aux" is the effects-signal mix.
- Line/mixdown—a combination found only in the Clarion unit. "Line" is for overdubbing. It lets you hear a mono mix of live signals and tape signals over headphones for selected tracks. Use "mixdown" to hear a stereo mix of all four tracks during mixdown.

Some units have no monitor-select switches. Instead, you must continually monitor the two-channel stereo monitor mix.

Headphone Volume Control

This controls the loudness of the headphones. All but the least expensive units have this feature.

Recorder Section of the Recorder-Mixer

The 4-track cassette deck built into the recorder-mixer also has many features to investigate. We covered recorder specifications and noise reduction in Chapter 4.

Overdubbing

A standard feature in all multitrack recorders, overdubbing entails recording a new track in sync with old tracks. When a musician overdubs, he or she listens through headphones to previously recorded tracks, plays a new part along with them, and records that new part on a blank track. Overdubbing provides many advantages. Since only one instrument or vocal is recorded at a time, unwanted sound from other instruments (leakage) is eliminated. The result is a clearer recording. Another advantage of overdubbing is that musicians who play more than one instrument can overdub so that the recording includes more instruments. A soloist can become a band by overdubbing all the parts him- or herself. A drawback of overdubbing is that you lose the emotional interaction and spontaneity that occurs among musicians when they all play together.

Synchronous Recording

This feature is used during overdubbing to keep prerecorded tracks synchronized with new parts that are being added "live." Prerecorded tracks are played back from the record head, rather than the playback head to keep the timing of the old and new parts in sync. Since all recorder-mixers combine the record and playback head into one, there are no sync problems.

Punch-In and -Out

With this feature, you can fix a mistake on a track without doing the whole track over. As the track is playing, you *punch-in* the record button at the appropriate spot, and the musician plays a corrected version that is recorded over the previous performance. When the musician has finished playing the corrected part, you *punch-out* of record mode so that the rest of the track is not erased. All recorder-mixers accept a footswitch so that the musician can punch in and out while performing.

Tape Counter

Analog counters are found in inexpensive machines and the slightly more accurate digital counters are found in more expensive machines.

Return-to-Zero

This function is also called *zero stop, autolocate,* or *memory rewind.* When you enable return-to-zero and hit rewind, the recorder automatically rewinds to a preset point marked 000 on the tape counter. This is useful for practice of punch-ins and mixes.

Some high-end units can repeatedly shuttle between two preset points, say, at the beginning and end of an overdubbed section.

Tape-Speed Options

A cassette recorder that operates at 1⅞ inches per second (ips) is compatible with commercial prerecorded cassettes, so you can play them on your recorder-mixer. A 3¾ ips recorder will not play standard prerecorded cassettes correctly, and uses tape twice as fast, but it provides better sound quality (extended high-frequency response, less tape hiss and less wow and flutter). Some recorders offer selectable tape speeds.

Pitch Control

This varies the speed of the cassette recorder, which lets you adjust the pitch of previously recorded tracks to match the tuning of new instruments to be added. Pitch variation ranges from ±10% to ±15% among different recorder models.

By understanding recorder-mixer features, you can choose a unit that suits your needs—and your budget. Recorder-mixer operation is covered in Chapter 9. But before using one in a session, we need to learn about signal processors, microphone techniques, and tape recorders.

6 Signal Processors

You can create special sonic effects, improve sound quality, and even enhance the music you record through the use of signal processors. Usually external to the mixing console, this outboard equipment takes a signal fed from the console and modifies it in a controlled way. Then the modified signal is returned to the console for routing to the appropriate channels. The result is a recording that sounds more like a professional production and less like a bland documentation.

Used on all pop records, effects such as reverberation, echo, and chorus add spaciousness and excitement. The signal processors that produce such effects cost $125 and up. They're a worthwhile investment if you want to produce a commercial sound. Recordings that include the natural room acoustics, such as recordings of classical-music ensembles and some folk groups, need no signal processing.

This chapter describes the most popular signal processors and tells how to use them.

The Equalizer

An equalizer (usually in the mixing console) is a sophisticated tone control, something like the bass and treble controls on a hi-fi set. Equalization affects tone quality by boosting or cutting selected frequency bands. That is, it alters the frequency response. EQ (pronounced "E.Q.") is studio jargon for equalization.

To understand how an equalizer works, we first need to know what a spectrum is. A musical instrument produces a wide range of frequencies, even when a single note is sounded. These frequencies, including the fundamental and harmonics, are the *spectrum* of the instrument, and they give the instrument a distinctive tone quality, or timbre.

Tone quality is affected by a change in the level of any portion of the spectrum. An equalizer raises or lowers the level of a particular range of frequencies (a frequency band), and so it controls the tone quality. For example, a boost (a level increase) at 10 kHz makes many instruments sound bright and crisp. A cut at the same frequency dulls the sound.

Types of Equalizers

Let's examine several types of equalizers, ranging from simple to complex.

Bass and Treble Control

The most basic equalizer is a bass and treble control. Its effect on frequency response is shown in Figure 6-1. Typically, such a device provides up to 15 dB of boost or cut at 100 Hz (for the low-frequency equalization knob) and at 10 kHz (for the high-frequency equalization knob).

Multiple-Frequency Equalizer

This equalizer allows boost or cut at several preset frequencies, as shown in Figure 6-2.

Sweepable Equalizer

This lets you tune in the exact frequency range to boost or cut, as shown in Figure 6-3.

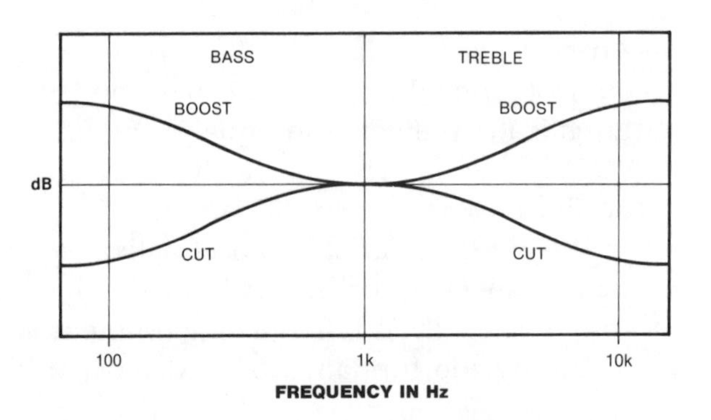

Figure 6-1. The effects of bass and treble controls.

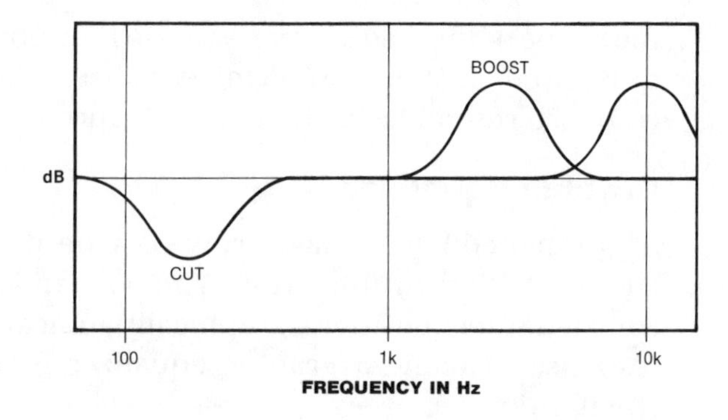

Figure 6-2. Multiple-frequency equalization.

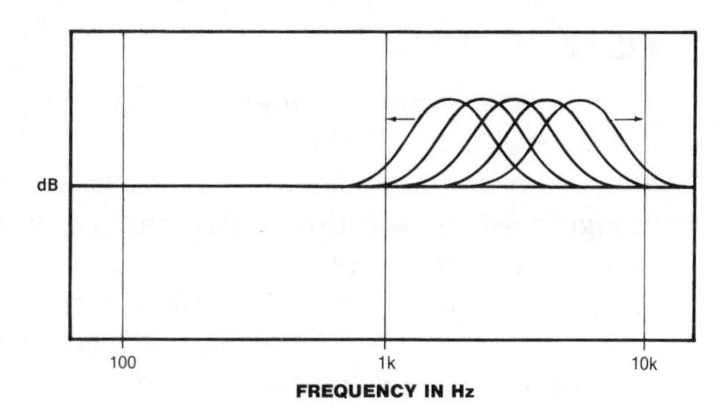

Figure 6-3. Sweepable equalization.

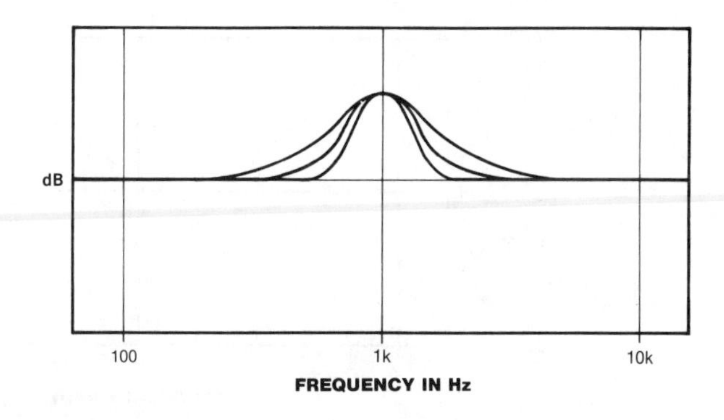

Figure 6-4. Varying the bandwidth of a parametric equalizer.

Parametric Equalizer

Most complex is a parametric equalizer, which allows continuous adjustment of frequency, level, and *bandwidth*—the range of frequencies affected. Figure 6-4 shows how a parametric equalizer varies the bandwidth of the boosted portion of the spectrum.

Peaking or Shelving

An equalizer is described as peaking or shelving depending on how it affects frequency response. With a *peaking* equalizer set for a boost, the shape of the frequency response resembles a hill or peak, as shown in Figure 6-5. With a *shelving* equalizer, the shape of the frequency response resembles a shelf, as in Figure 6-6.

Graphic Equalizer

A graphic equalizer has a row of slide potentiometers dividing the audible spectrum into from 5 to 31 bands. When the controls are adjusted, their positions graphically indicate the resulting frequency response. Usually, a graphic equalizer is used for monitor-speaker equalization.

Filter

A filter is a form of equalizer that sharply rejects (*attenuates*) frequencies above or below a certain frequency. For example, a 10-kHz lowpass, or high-cut, filter removes frequencies above 10 kHz (as shown in Figure 6-7). This reduces hiss without affecting tone quality as much as a gradual treble rolloff would. A 100-Hz highpass, or low-cut, filter attenuates frequencies below 100 Hz, reducing rumble from air conditioning and trucks. Filtering out frequencies above and below the spectral range of a musical instrument reduces leakage at those frequencies.

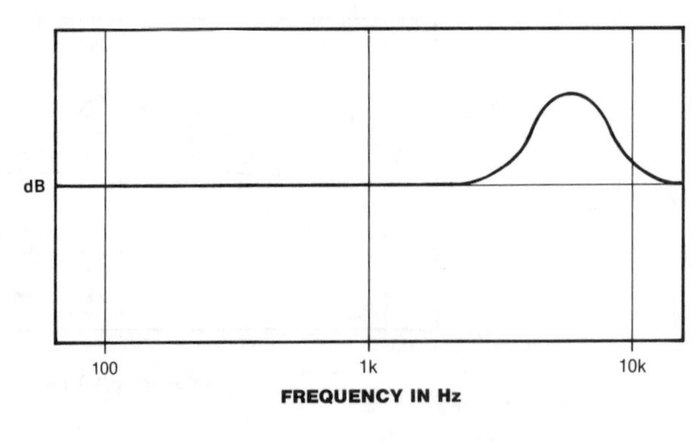

Figure 6-5. Peaking equalization at 7 kHz.

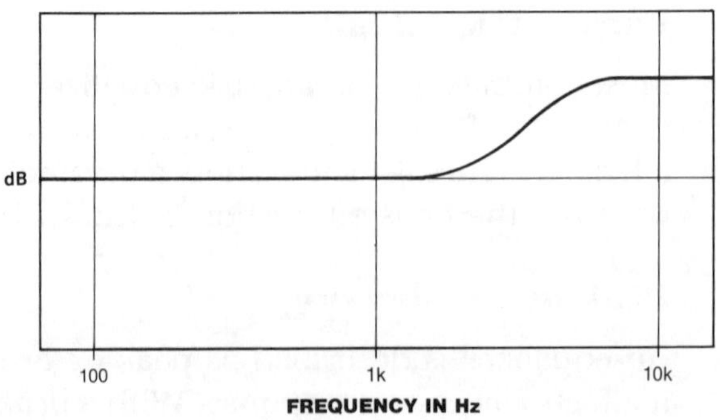

Figure 6-6. Shelving equalization at 7 kHz.

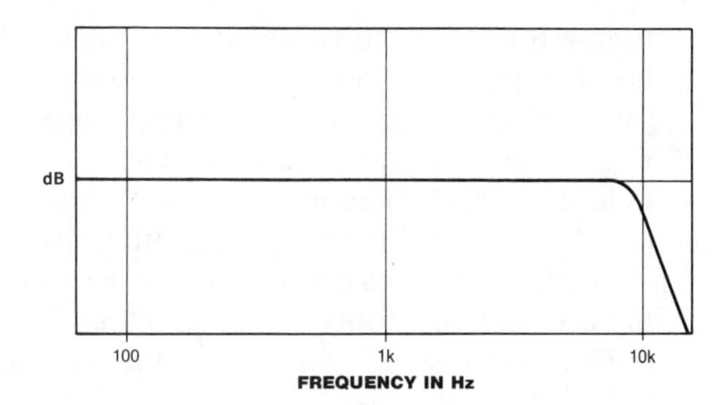

Figure 6-7. 10-kHz lowpass filter (−3 dB at 10 kHz).

Setting Equalization

One way to set an equalizer is to first set it to the approximate frequency range you need to work on (you'll soon know where by experience). Next, apply full boost so the effect is easily audible. Finally, fine-tune the frequency and amount of boost or cut until the tonal balance is the way you like it.

For example, if a close-miked vocal sounds unnaturally bassy, turn down the low-frequency equalization knob (say, 100 Hz), adjusting the amount of cut for the desired tonal balance.

If you hear a strong coloration in the tone quality of an instrument, set a sweepable equalizer for extreme boost to find the frequency range matching the coloration. Then cut that range by an amount that sounds right.

It's instructive to spend some time using a graphic equalizer. Play a musical program through it to hear the tonal effect of each frequency band. Also play individual tracks of different instruments through it, and note how a boost at a certain frequency has a different effect on each instrument. Then you'll know what frequency to boost or cut in order to correct a tonal coloration. With a little experimentation, you'll get a better idea what knob to turn on any equalizer to get a "woody" sound or a "brassy" sound. At the end of this chapter is a table on sound quality descriptions. Among other things, it suggests equalizer settings to achieve various sonic effects.

When to Equalize

Should equalization be applied during recording or during mixdown? If you're mixing the instruments live to two-track as the music is performed, there is no separate mixdown session, so you must apply equalization during recording. If you're assigning several instruments to one track, you must equalize these instruments during recording because you can't equalize them individually during mixdown. The same restriction is true for adding reverberation or other effects to instruments assigned to the same track. If you assign each instrument to its own track, however, the usual practice is to record *flat* (without equalization) and then equalize the track during mixdown.

If you use a bass cut or treble boost, you can obtain a better signal-to-noise ratio by applying equalization during recording, rather than during mixdown. If you use a treble cut, applying it during mixdown will reduce tape hiss.

Uses of Equalization

The following is a partial list of how equalization is used.

Improving Tone Quality

Equalization can make an instrument sound better tonally. For example, you might rolloff (turn down) the high frequencies on a singer to reduce sibilance or on a direct-recorded electric guitar to take the "edge" off the sound. As another example, boosting 100 Hz on a floor tom gives a fuller sound, or cutting around 250 Hz on a bass guitar aids clarity. The frequency response and placement of each microphone affect tone quality as well.

Special Production Effects

Extreme equalization reduces fidelity, but it also can make interesting sound effects. Sharply rolling off the lows and highs on a voice, for instance, gives it a "telephone" sound. An extreme boost at 5 kHz can accent the impact of a snare drum.

Helping a Track Stand Out

A recorded track of an instrument heard by itself may sound very clear, but when it's mixed with other tracks, the clarity may disappear. Certain frequencies of the instrument can be covered up or masked by frequencies produced by other instruments. A boost in the presence range, say 1.5 kHz to 6 kHz, can help restore presence and clarity. Vocals typically are boosted in this range to help them stand out against an instrumental background.

Compensating for Response Deficiencies

The microphones, tape recorder, monitor speakers, and the mixing board itself may not have a flat frequency response. Equalization can partly compensate for these deficiencies. If a microphone has a gradual high-frequency rolloff, for example, a high-frequency boost on the console may help restore a flat response. On the other hand, if a microphone "dies" above a certain frequency, no amount of boost can help it. Some directional microphones have *proximity effect*—a bass boost when used up close. A bass rolloff on the console can compensate for this boost.

Compensating for Microphone Placement

Often you must place a microphone very close to an instrument in order to reject background sounds and leakage. Unfortunately, a closely placed microphone tends to emphasize the part of the instrument that the microphone is near; the tone quality picked up may not be the same as that of the instrument as a whole. Equalization can partly compensate for this effect. For example, a guitar miked next to the sound hole sounds bassy because the sound hole radiates strong low frequencies, but a complementary low-frequency rolloff on the console can restore the natural tonal balance.

Reducing Noise and Leakage

By filtering out frequencies above and below the spectral range of an instrument, you can reject noise and leakage at those frequencies. For instance, a kick drum has little or no output above 5 kHz, so you can filter out highs above 5 kHz on the kick drum to reduce cymbal leakage. If you do this filtering during mixdown, it will also reduce tape hiss. Filtering out frequencies below 100 Hz on most instruments reduces air-conditioning rumble and muddy bass.

Compensating for the Fletcher-Munson Effect

As discovered by Fletcher and Munson, the ear is less sensitive to bass and treble at low volumes than at high volumes. So, when you record a very loud instrument and play it back at a lower level, the recording might lack bass and treble. To restore these, you may need to boost the lows (around 100 Hz) and the highs (around 4 kHz) when recording loud rock groups, for example. The louder the group, the more boost you'll need. As an alternative, use cardioid microphones with proximity effect (for bass boost) and a presence peak (for treble boost).

Making a Pleasing Blend

When several instruments are heard together, they sometimes "crowd," or overlap, each other in the frequency spectrum. That is, it may be difficult to distinguish the instruments by tonal differences. But by equalizing the various instruments at differing frequencies, you can make their timbres distinct, which results in a more pleasing blend. This procedure also evens out the contribution of each frequency band to the total spectrum, yielding a mix that is well-balanced tonally.

The Compressor

This device acts like an automatic volume control, turning down the volume if the signal gets too loud. The following example shows why a compressor is necessary.

When you record a vocalist sometimes he sings too softly and gets buried in the mix; other times he hits loud notes, saturating the tape. Or he may move toward and away from the microphone while singing, so that the average recording level fluctuates.

To control this problem, you can *ride gain* on the vocalist—turn down the gain (amplification) when the vocal is too loud; turn it up when the vocal is too quiet. But, since it's hard to anticipate these changes, it's better to use a *compressor,* an amplifier that performs the same function automatically. It reduces the gain when the input signal exceeds a preset level (the *threshold*). The greater the input level, the less the gain. As a result, quiet passages are made louder and loud passages are made softer and so the dynamic range is reduced (as shown in Figure 6-8).

Compression keeps the level of vocals or instruments more constant, making them easier to hear throughout the mix and preventing loud notes that may saturate the tape. With extreme control settings, a compressor also can be used for a special effect—say, to make drums

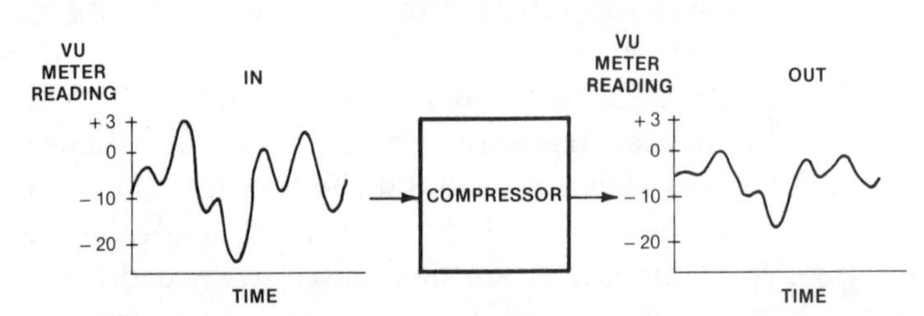

Figure 6-8. Compression.

sound "fatter." In professional studios, compression is nearly always applied to vocals, often to bass guitar and drums, and occasionally to piano and lead guitar.

You might be able to get by without a compressor if the vocalist uses proper mic technique: he or she should back away from the mic on loud notes, and come in close on soft notes. To tell whether you need a compressor, listen to your finished mix. If you can understand the words throughout the song without the loud vocal notes being too loud, you probably can do without a compressor.

Using a Compressor

Normally, you compress individual instruments or tracks rather than the entire mix. That procedure makes the effect less audible because it applies compression only to those instruments needing it.

Compressing instrument signals during recording improves the signal-to-noise ratio of the tape tracks, but it forces you to decide on compressor settings during the recording session. Compressing tape tracks during mixdown allows you to change the settings at will, but it can make tape hiss audible by raising the gain during quiet passages.

Several controls on the compressor need careful adjustment. Some of the following parameters are preset internally on various models.

Compression Ratio or Slope

This is the ratio of the change in input level to the change in output level. For example, a 2:1 ratio means that for every 2-dB change in input level, the output changes 1 dB. A 20-dB change in input level would result in a 10-dB change in the output, and so on. Ratio settings from 1.5:1 to 4:1 are typical.

A "soft knee" or "over easy" characteristic is a low compression ratio for low-level signals and a high compression ratio for high-level signals. Some manufacturers say that this characteristic sounds more natural than a constant compression ratio.

Gain Reduction

This is the number of decibels that the gain or level is reduced by the compressor. It varies with the input level. You set the ratio and threshold controls so that the gain is reduced on loud notes by an amount that sounds right, or looks right on the gain-reduction meter.

Attack Time

The attack-time setting controls how fast the gain reduction occurs in response to a musical attack. Typical attack times range from 0.25 to 10 milliseconds (ms). Some compressors adjust attack time automatically to suit the program material; others have a factory-set attack time. The longer the attack time, the larger are the peaks that pass before gain reduction occurs. Thus, a long attack time sounds punchy; a short attack time reduces punch by softening the attack.

Release Time

The release-time or recovery-time control affects how fast the gain returns to its normal value after a loud passage. As the gain returns to normal, noise increases along with the signal, resulting in a "pumping" or "breathing" sound. Release time, which is adjustable from about 50 ms to several seconds, usually is set for the least objectionable effect, depending on program material. Shorter release times make the compressor follow faster dynamic changes in the music and keep the average level higher. In some units, the release time is adjusted automatically or is factory-set to a useful value. Release time for bass instruments must be longer than about 0.4 second to prevent harmonic distortion.

Threshold

This is the input level above which compression occurs. Set the threshold high (near 0 VU) to compress only the loudest notes; set it low (−10 or −20 VU) to bring up quiet passages as well as softening loud ones. If the compressor has a fixed threshold, the input level control is used to adjust the amount of compression.

Output-Level Control

This control sets the signal strength coming out of the compressor to the proper level for your mixer input. Some units automatically maintain a constant output level when other controls are varied.

Connecting a Compressor

You connect a compressor in series with the signal you want to compress, in one of the following ways.

- If you want to compress a single instrument or voice, and your mixer has access jacks, locate the input module that controls the source you want to compress. Connect the compressor between the access jacks.
- If you want to compress the signal of a console output channel (bus) while recording, locate the bus output containing the signal of the source(s) you want to compress. Connect this output to the compressor input and connect the compressor output to the desired tape-track input.
- If you want to compress a particular tape track during mixdown, connect the tape-track output to the compressor input and connect the compressor output to the console line input normally used for that track. Alternatively, locate that track's input module in the console and connect the compressor between the access jacks for that input module.

The Noise Gate

A *noise gate*, or *expander*, is patched between a tape-track output and a mixing-board line input. The gate acts like an on-off switch to eliminate noises during pauses in an audio signal. It does this by reducing the gain when the input level falls below a preset threshold. That is, when an instrument momentarily stops playing, the signal level is low enough so that the noise gate turns off—eliminating any noise and leakage during the pause (as shown in Figure 6-9).

The noise gate helps to clean up drum tracks by removing leakage between beats. It also can be used for special effect to shorten the decay time of the drums, producing a very tight sound. Sometimes during a mixdown, a gate is used on each output of the multitrack recorder to reduce tape hiss. The noise-gate threshold should be set high enough to chop off tape hiss during pauses, but low enough to

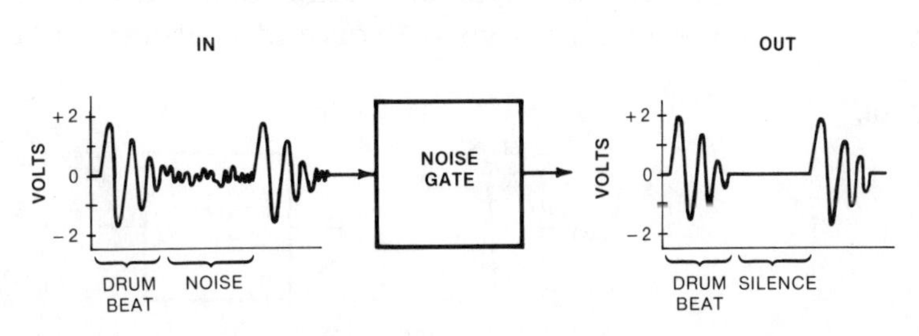

Figure 6-9.
Noise gating.

avoid removing any program material (unless that is the desired effect). The release time should be very fast for drums and longer for more sustained instruments.

Adequate demo tapes can be made without gating. But for those who want a little tighter, cleaner sound, gates are worthwhile.

The Delay Unit

By delaying a signal, a wide variety of special effects can be created, including echo, multiple echo, doubling, chorus, and flanging. Before explaining these, let's look at the device that performs all these wonders: a delay unit.

A delay unit or digital delay accepts an input signal, holds it in an electronic memory, then plays it back after a short delay—about 1 ms to 1 second (illustrated in Figure 6-10). Delay is the time interval between the input signal and its repetition at the output of the delay device.

Delay-Unit Specifications

The bandwidth of the delay unit is the frequency range or upper frequency limit of the delayed signal. A 12-kHz bandwidth is good, 16 kHz is excellent, and 20 kHz is icing on the cake.

The signal-to-noise ratio (S/N) of the delay unit is the ratio in decibels between the delayed signal's level and the noise level. In general, the longer the delay, the poorer the S/N. A ratio of 70 dB is considered fair for delay units, 80 dB is good, and 90 dB is very good.

Echo

Delay by itself is not an audible effect. If we delay the incoming signal, however, by between 50 ms and 1 second, and combine the undelayed

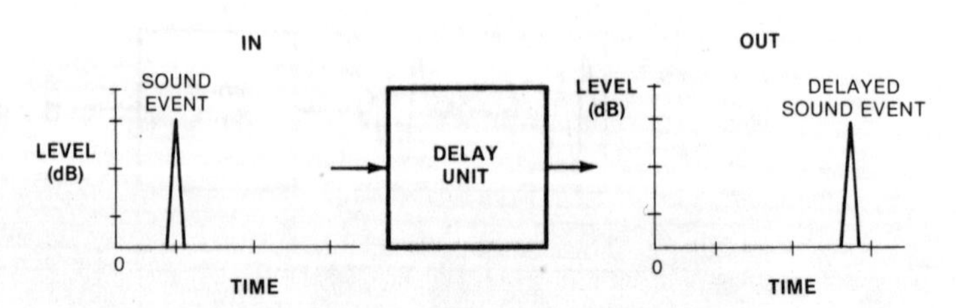

Figure 6-10. Delay.

and delayed signals, we hear two distinct sounds: a signal and its repetition. The delayed repetition of a sound is an echo and it is illustrated in Figure 6-11. Echoes occur naturally when sound waves travel to a room surface, bounce off, and return later to the listener—repeating the original sound. A delay unit can mimic this effect.

The delayed and direct (undelayed) signals can be combined in the delay device by setting its direct/delay mix control partway up. To use this method, connect the delay unit in-line with the track you want echoed.

There's another way to mix the delayed signal with the undelayed signal to make an echo: through the mixing console. Proceed as follows:

1. Plug the signal to be delayed into your mixing console.
2. Set the direct/delay mix control on the delay unit all the way to "delay." This control might be labeled "dry/wet"; if so, set it to "wet."
3. On the mixing console, locate the *aux-send* output (also called "effects send"). Connect it to the delay unit input.
4. Connect the delay unit's output to the console *bus-in* connector (also called "effects return").
5. Set the bus-in knob about halfway up.
6. Locate the input module where you plugged in the signal to be delayed. The fader in this module controls the level of the undelayed sound; the aux-send knob in this module controls the level of the delayed sound. Mix the two in the desired proportion.

Slap Echo

A delay around 50 to 200 ms results in a slap echo or slapback echo—often used in 1950s rock 'n' roll tunes, and still used today.

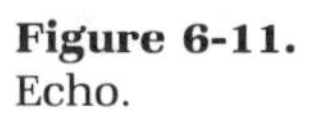

Figure 6-11.
Echo.

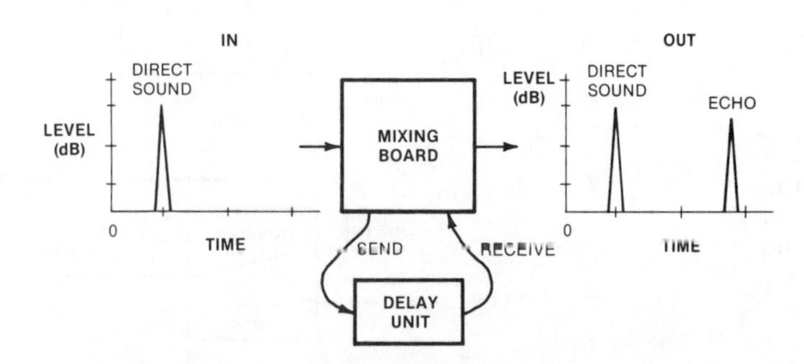

Multiple Echo

Many delay units have a feature called *recirculation* or *regeneration.* With this feature, some of the delayed output is fed back into the input, so that the signal is re-delayed many times. This creates a *multiple echo*—several repetitions that are evenly spaced in time (illustrated in Figure 6-12).

You recirculate the delayed sound by turning up the recirculation control on the delay unit. If your unit has no such control, you can set up an external recirculation system. Patch (connect) its output into a spare console line input, and turn up that input's effects-send control (feeding the delay device) for the desired effect, as shown in Figure 6-13. The higher the recirculation level, the longer the echoes last.

Multiple echo is most musical if you set the delay time to create an echo rhythm that fits the tempo of the song. A slowly repeating echo—say, 0.5 second between repeats—gives an outer-space or haunted-house effect.

Doubling

If the delay is set around 15 to 35 ms, the effect is called *doubling* or *automatic double tracking (ADT).* This gives an instrument or voice a fatter, stronger sound, especially if the original signal is panned left and the delayed signal is panned right, or vice versa. The short delays used in doubling create a sound similar to early sound reflections in a studio—thus they add a sense of "air," or ambience, to close-miked instruments that would otherwise sound too dead.

Doubling a vocal can be done without a delay unit in this way:

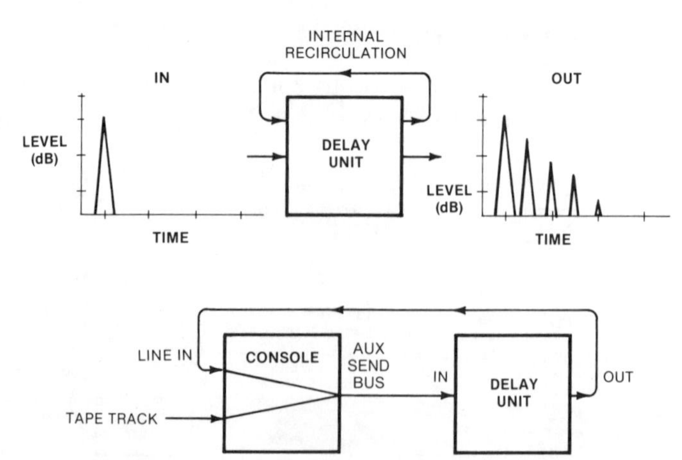

Figure 6-12. Multiple echo.

Figure 6-13. Setup for multiple echo using external recirculation.

1. Record a vocal part.
2. Rewind the tape to the beginning.
3. Set the tape recorder in sync mode so the singer can hear the performance just recorded and sing along with it.
4. On an unused track, rerecord the singer in another performance of the same vocal part.
5. Rewind the tape and play back the two synchronized performances. The doubled vocal sounds fuller than a single vocal track. (This method was often used by the Beatles.)

Chorus

If the delay used in doubling is *modulated* (slowly varied), this produces a wavy or shimmering effect called *chorus.* Feeding some of the delayed output back into the input (regeneration) adds extra fullness. Chorus can make a single voice sound like a chorus of voices singing in unison, or give a lead guitar a spacious, "singing" quality. Stereo chorus is an especially beautiful effect.

Flanging

If the delay is set around 0 to 20 ms, the ear is usually unable to resolve the direct and delayed signals into two separate sounds. Instead, a single sound with an unusual frequency response is heard. Due to phase cancellations of the direct and delayed signals combined, a series of peaks and dips results in the net frequency response called a *comb-filter effect,* shown in Figure 6-14. It gives a colored, filtered tone quality. The shorter the delay, the farther apart are the peaks and dips spaced in frequency.

In a flanger (or in a digital delay set to flanging mode), the delay is automatically varied (swept) from about 0 to 20 ms. This causes the comb-filter nulls to sweep up and down the spectrum. The resulting

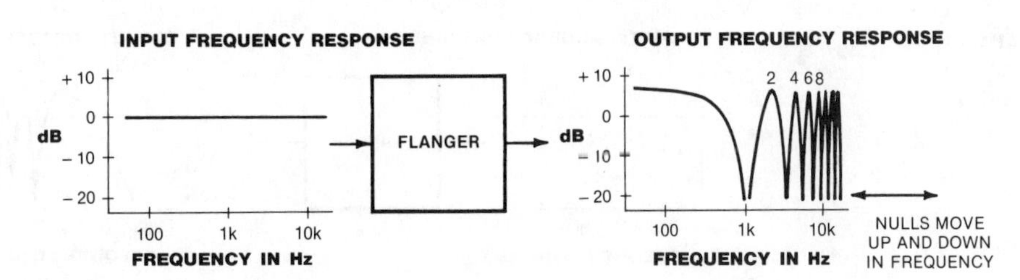

Figure 6-14. Flanging (or positive flanging).

sound quality is hollow, swishing, and ethereal, as if the music were playing through a variable-length pipe. Flanging is applied most effectively to broad-band signals such as cymbals but can be used on any instrument.

Used most often in psychedelic music, flanging can be heard on many Jimi Hendrix recordings, as well as the song "Itchycoo Park" by the Small Faces.

Positive and Negative Flanging

Positive flanging refers to flanging in which the delayed signal is the same polarity as the direct signal. With *negative flanging,* the delayed signal is opposite in polarity to the direct signal, which creates a stronger effect. The low frequencies are canceled (the bass rolls off), and the "knee" of the bass rolloff moves up and down the spectrum as the delay is varied. The high frequencies are still comb-filtered, as shown in Figure 6-15. Negative flanging makes the music sound like it is turning inside out!

Resonant Flanging

By feeding some of the output of the flanger back into the input, the peaks and dips are reinforced, creating a powerful science-fiction effect called *resonant flanging.*

The Reverberation Unit

A *reverberation unit* adds a sense of room acoustics, ambience, or of space surrounding the instruments and voices. It's a necessity and should be the first signal processor you buy. Here's how it works.

Acoustic *reverberation* is a series of multiple sound reflections that makes the original sound persist and gradually die away or decay. These reflections tell your ear that you're listening in a large or hard-surfaced room. An artificial reverberation unit simulates the sound of

Figure 6-15. Negative flanging.

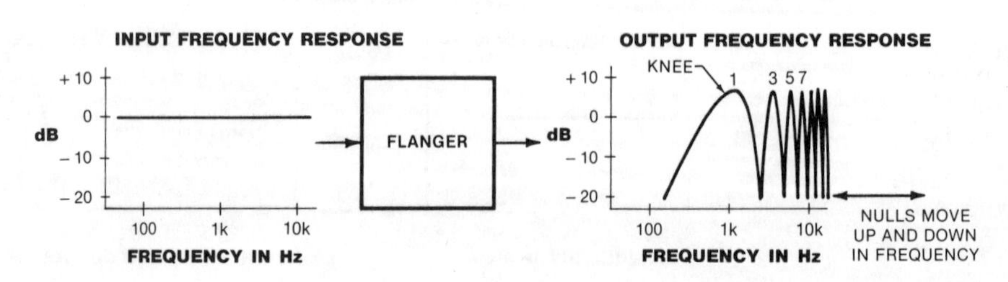

an acoustic environment—a club, auditorium, or concert hall—by generating random multiple echoes that are too numerous and rapid for the ear to resolve. Reverberation is illustrated in Figure 6-16.

Note that reverberation is sometimes called "echo," although an echo is a distinct repetition of a sound, rather than a continuous decay of sound. Delay units make echoes; reverb units make reverberation.

A digital reverb unit (Figures 6-17 and 6-18) offers a wide variety of reverberation patterns. It uses sophisticated programs that mimic the early and late sound-reflection patterns of rooms of various size. It even can duplicate the bright sound of a metal-foil plate, which used to be the most popular type of reverb device in professional studios. Some units let you adjust tone quality, decay time, and other parameters. Unnatural effects are available, such as nonlinear decay, reverse reverberation which builds up before decaying, or gated reverberation.

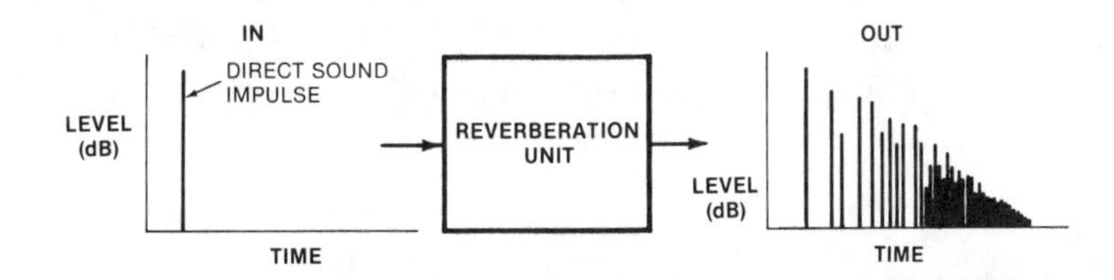

Figure 6-16. Reverberation.

Figure 6-17. The Yamaha REV5 Digital Reverberator *(Courtesy of Yamaha Electronic Corporation USA)*

Figure 6-18. The Alesis Microverb II digital reverberator: a low-cost, professional-quality unit with a variety of reverberation programs *(Courtesy of Alesis Corp.)*

With *gated reverberation,* the reverberation cuts off shortly after a note is hit. It's often used on snare drums. A good example is the song "You Can Call Me Al" on Paul Simon's "Graceland" recording. To produce this effect, you either use a reverberation unit with a gated-reverb program, or feed the reverberation-return signal through a noise gate set to cut off the end of the reverberant "tail."

Another feature in some units is *predelay* (or *prereverb delay*). This is a short delay (say, 30 to 100 ms) before the reverberation to simulate the delay that occurs in real rooms before the onset of reverberation. The longer the predelay, the greater the sense of room size. If your reverb unit does not have predelay built in, you can create it by connecting a delay unit between the console aux-send output and the reverb-unit input.

To connect a reverb unit to your mixer, run a cable from the mixer aux-send output to the reverb input and run a cable (two for stereo) from the reverb outputs to the mixer effects returns or bus inputs. Turn the bus-in knob (if any) to about halfway, then use the aux-send knobs to adjust the amount of reverb on each track.

The Enhancer

This device takes dull material and puts brilliance, sparkle, and clarity back into it—without adding noise as EQ can. It works either by adding slight distortion (as in the Aphex Aural Exciter) or by boosting the treble when the signal has high-frequency content (as in the Alesis Micro Enhancer and the Barcus Berry Electronics 402 Sonic Maximizer). This last device also divides the frequency range into three bands: the lows are delayed about 1.5 ms, the midrange is delayed about 0.5 ms, and the highs are delayed only a few microseconds. Thus, harmonics and fundamentals are realigned in time, which, it is claimed, aids clarity.

The Octave Divider

This unit accepts a signal from a bass guitar and provides deep, growling bass notes one or two octaves below the pitch of the bass guitar. It does this by dividing the incoming fundamental frequency by 2 or

4; if you put 82 Hz in, you get 41 Hz out. This effect is becoming popular in contemporary recordings.

Summary of Signal-Processor Effects

Here's a brief definition of each effect we discussed:

- Equalization (EQ)—boosts or cuts selected frequency bands to alter tone quality (bass, midrange, and treble).
- Compression—reduces dynamic range.
- Noise Gating—removes noise and leakage during pauses.
- Echo—repeats a sound after a delay of 50 ms to 1 second.
- Multiple Echo—recirculates an echo (several repetitions).
- Doubling—delays a signal from 15 to 35 ms for ambience and fullness.
- Chorus—varies or sweeps the delay used in doubling (possibly with regeneration). This produces a wavy, shimmering effect.
- Stereo chorus—same as chorus but in stereo for a spatially "fat" or full effect.
- Flanging—sweeps a delay from 0 to 20 ms. This gives a hollow, swishing, ethereal effect, as though a jet were passing overhead.
- Reverberation—simulates room acoustics and adds spaciousness.
- Enhancer—adds brilliance (treble) either by adding distortion or by boosting high frequencies when the music contains high frequencies.
- Octave divider—produces notes one and two octaves below the note played for a deep-bass effect.

Rock music uses all of these effects, although flanging is growing dated. The guitar in rock, jazz, or new age music often benefits from chorus. Reverberation is useful for all types of music. The same is true of the enhancer if the recording lacks brilliance. Compression is needed mainly for vocals.

Recently available are digital programmable processors (multiprocessors), such as that shown in Figure 6-19. These multipurpose units have several preset effects that you can modify and save if you wish.

Figure 6-19. The Yamaha REX50 Digital Multi-Effects Processor. *(Courtesy of Yamaha Music Corporation)*

Some can combine several effects at once, and many are MIDI-controllable.

The first signal processors a home-studio owner should purchase are a good reverberation unit and a compressor. These devices are practically indispensable in producing a commercial sound.

Special effects help define the characteristic recorded sound of an era. The '50s had the "tube sound" and slap echo; the '60s used "fuzz," "wah-wah," and flanging; many recordings made in the '70s sounded rather dead; the early '80s introduced delay units and enhancers; and today's sound emphasizes MIDI keyboards, drum machines, and digital reverb effects. You make recordings that sound contemporary by using the latest effects and musical instruments on the market and by inventing new sounds. It's the creative usage of these effects—combining and using them in unusual ways—that leads to attention-grabbing recordings.

Sound-Quality Descriptions

All these signal processors produce sonic effects that may be difficult to describe in technical terms. For example, what equalization should be used to get a fat sound or a thin sound? What physical conditions may be causing a muddy or a metallic sound? In general, which knob do you turn to achieve a certain sonic effect?

Table 6-1 answers these questions. It translates audio-engineering terms (such as equalization settings) into subjective descriptions of sound quality. Note: These definitions are not universally agreed upon, but they are probably the most common meanings. The positive terms apply when you like the effect; the negative terms apply when you don't!

Table 6-1. Translation of audio-engineering terms into subjective descriptions of sound quality.

Low-Frequency Boost (below about 500 Hz)

Positive	*Negative*
Powerful (under 200 Hz)	Muddy
Ballsy (under 200 Hz)	Tubby (200–300 Hz)
Heavy (under 200 Hz)	Thumpy
Fat	Boomy
Thick	Barrel-like
Woody (200–400 Hz)	Woody (200–400 Hz)
Warm	
Robust	
Mellow	
Full	

Flat, Extended, Low Frequencies

Positive	*Negative*
Full	Rumbly
Full-bodied	
Rich	
Solid	
Natural	

Low-Frequency Rolloff

Positive	*Negative*
Clean	Thin
	Cold, cool
	Tinny
	Anemic

continued

Table 6-1.
(cont.)

Mid-Frequency Boost (5-kHz area for most instruments, 1.5–2.5 kHz for bass instruments.)	
Positive	*Negative*
Present (Presence)	Hollow, muffled (500 Hz)
Punchy	Muddy, horn-like (500 Hz)
Edgy	Edgy (3–7 kHz)
	"Aw" sound (500–800 Hz)
Clear	Tinny, telephone-like (1 kHz)
Intelligible	"Er" sound (1.5 kHz)
Articulate	Nasal, honky (500 Hz to 3 kHz)
Defined	Hard (2–4 kHz)
Projected (2–3 kHz)	Harsh, strident, piercing (2–5 kHz)
Forward (2–3 kHz)	Metallic (3–5 kHz, especially 3 kHz)
	Twangy (3 kHz)
	Sibilant (4–7 kHz)
Flat Mid-Frequencies	
Positive	*Negative*
Natural	No punch
Neutral	"Flat" (lacking character or color)
Smooth	
Musical	
Mid-Frequency Dip	
Positive	*Negative*
Mellow	Hollow (500–1000 Hz)
	Disembodied (500–1000 Hz)
	Muffled (5 kHz)
	Muddy (5 kHz)
High-Frequency Boost (above about 7 kHz)	
Positive	*Negative*
Trebly	Trebly
Sizzly (cymbals)	String Noise
Bright	Sizzly (voice)
Crisp	Edgy
Articulate	Glassy
Etched	"Essy" Sibilant
Hot	Steely

Flat, Extended High Frequencies

Positive	*Negative*
Open	Too detailed
Airy	Too close
Transparent	
Clear	
Natural	
Neutral	
Smooth	
Effortless	
Detailed	

High-Frequency Rolloff

Positive	*Negative*
Mellow	Dull
Round	Restricted
Smooth	Muffled
Easy-on-the-ears	Veiled
Concert hall-like	Muddy
Dark	Distant

Overall Response

Positive (all flat response)	*Negative*
Natural	Rough, peaky, harsh, colored (nonflat, peaks and dips)
Accurate	Phasey (sharp dips)
Neutral	Cheap (narrow-band)
Smooth	Flat (lacking character—too neutral)
Transparent	
Effortless	
Musical	
Uncolored	
Liquid	

continued

Table 6-1.
(cont.)

Reverberation or Leakage

Too Little	*Well-Controlled*	*Pleasant*	*Too Much*
Sterile	Clean	Warm	Echoey
Dry	Tight	Rich	Bathroom-sound
Dead		Sumptuous	Muddy
Muffled		Airy	Loose
Thin		Having depth	Washed-out
		"Live"	Barrel-like
		Spacious	Cavernous
		Open	In another room
		Full	Distant
		Bright	Trashy

Noise and Distortion

Absent	*Present*
Clean	Veiled (mild distortion)
Clear	Hard
Smooth	Harsh
Open	Grainy
	Gritty
	Dirty (positive or negative)
	Distorted
	Fuzzy
	Sputtering
	Raunchy
	Noisy
	Hissy

Stereo Imaging

Sharp	*Diffuse*
Focused	Unfocused
Pinpointed	Vague
Easy-to-localize	Hard-to-localize
Fused	Smeared
Defined	Spread
Pan-potted	Directionless
	Spacious
	Hole-in-the-middle
	Phasey

	Fat
	Big

Relative Loudness in the Mix

Loud	*Quiet*
Up front	Distant
On top	Subtle
Present	In the background
Hot	Recessed
Forward	Lost
Dominating	Covered
Covering	

7 Microphone Techniques

Because mic technique is largely a matter of personal choice of mic and placement, there's no single correct microphone technique for any particular instrument or instrumentation. Still, many useful techniques have evolved and they are described in this chapter. After trying these techniques you should try making up your own techniques, too.

First a general principle: microphone placement affects the tonal balance of a recorded instrument. When you change the mic position, you change the tone quality. So, to find a good spot, try placing the microphone in various locations—and monitor the results—until you find one that sounds good to you.

Another principle: microphone placement also affects the sense of distance of a recorded instrument. Mike close to achieve a tight, present sound; mike farther away for a distant, spacious sound. The farther a microphone is from its sound source, the more the microphone picks up room acoustics, background noise, and leakage from other instruments.

Now let's explore some common miking techniques for specific instruments and vocals.

Electric Guitar

The electric guitar can be recorded in many ways, as shown in Figure 7-1.

- with a microphone in front of the guitar amp
- with a direct box
- both miked and direct
- through a signal processor

The style of music you're recording suggests the appropriate method. Miking the amp is best when you want a rough, raw sound including the vacuum-tube distortion and speaker coloration. Rock 'n' roll or Heavy Metal usually sound best with a miked amp. Recording with a direct box, on the other hand, sounds clean and clear and gives extended highs and lows. It might be the best method for recording quiet jazz or rhythm and blues (R&B). Regardless of style, however, use whatever sounds right for the particular song being recorded.

First work on reducing any hum heard through the guitar amp. Make sure a quality guitar cord is being used. Flip the polarity switch on the amp to the lowest-hum position. Set the guitar volume and treble controls up full. Have the guitarist move around to find a null spot in the room where hum disappears. You may want to try a noise gate to remove buzzes between guitar notes.

Miking the Amp

Small practice amplifiers are generally better for recording than large, noisy stage amplifiers. If you use a small amp, place it on a chair to avoid picking up reflections from the floor.

The most popular microphone choice for electric guitar is a cardioid moving-coil type with a presence peak in the frequency response (a boost around 5 kHz). Remember that moving-coil is often called "dynamic." The cardioid pattern reduces leakage, the moving-coil transducer withstands very loud sound without distorting, and the presence peak adds punch. Of course, you can use any mic that sounds good to you.

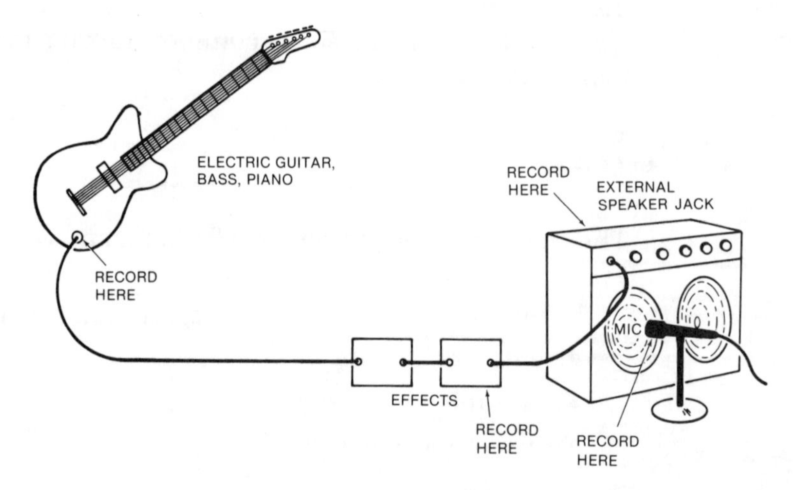

Figure 7-1. Recording an amplified-instrument system.

As a starting point, place the mic about 1″ to 1′ from the amp, aiming at the center of one of the speaker cones. The closer to the amp the mic is placed, the bassier the tone is and the less ambience and leakage are picked up. Placement in front of the center of the speaker cone produces a bright sound; off-center placement gives a more mellow sound and reduces amplifier hiss.

If you're overdubbing a lead guitar played through a huge stack of speakers in an acoustically live room, you may want to mike the amp at a distance.

Recording Direct

Now let's consider recording direct (also known as *direct injection* or *DI*). The electric guitar produces an electrical signal, so it can be plugged right into the mixing console—no microphone is needed. Since the mic and guitar amp are bypassed, the sound is clean and clear; it lacks the distortion and coloration of the amp. But remember, too, that amplifier distortion is desirable in some songs.

If you can use a short cable from guitar to mixer and if your mixer's inputs are phone jacks (¼″-diameter holes), then you can plug directly into the mixer. If your mixer has three-pin (XLR-type) mic inputs, you need a direct box (shown in Figure 7-2), which adapts the guitar signal to the mixer input.

Plug the guitar into the direct box and plug the direct box into a microphone input. Some direct boxes let you record off the amplifier's external-speaker jack to pick up distortion, which is often a desirable part of the sound. These boxes often include a lowpass filter to simulate the frequency response of the amp speaker. The direct box should have a ground-lift switch to prevent ground loops and hum. Set it to the lowest-hum position (usually lifted).

An inexpensive direct-connection cable was shown in Figure 2-5. This cable connects the amplifier speaker jack and a mixer mic input.

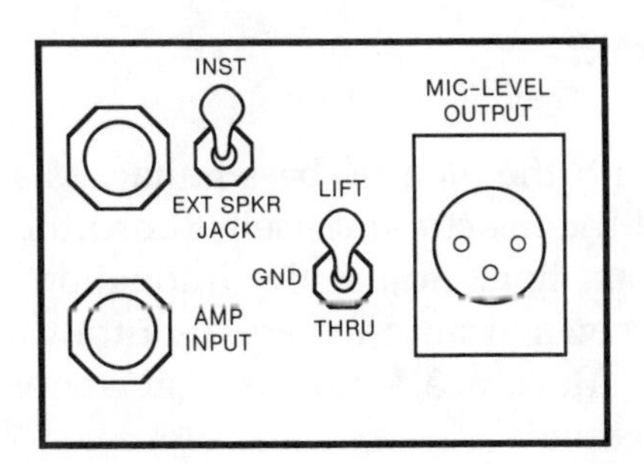

Figure 7-2.
A typical direct box.

Recording Direct from the Guitarist's Effects Boxes

If you want to record the guitarist's special effects but not the guitar-amplifier distortion, connect the output of the effects boxes to the direct-box input. Some engineers like to record a combination of direct and miked sound.

Electric-Guitar Studio Effects

You'll often want a thick or spacious lead-guitar sound. One way to get it is to send the guitar signal through a stereo chorus. To save a track during recording, record the guitar in mono on a single track and add the stereo chorus during mixdown. Connect the guitar-track output to the chorus-unit input; connect the two chorus outputs to your mixer line inputs.

Another way to thicken the sound is to double the guitar. Have the player rerecord the same part in sync with the original part. Pan the original part left and pan the doubled part right. Or "Y" the guitar cord to feed the guitar amp recorded on the left channel and a Leslie organ speaker recorded on the right channel. You'll hear a spacious, swirling sound. Reverberation and extreme compression are also popular effects. There are guitar-level signal processors (such as The Rockman by Scholz Research and Development, or the Pocket Rock-It by CB Labs) that add such effects to guitar as distortion, equalization, chorus, and compression. Simply by plugging the electric guitar straight into the Rockman, adjusting the switches for the desired sound, and recording the signal directly from the Rockman, you get a fully produced sound with a minimum of effort. The Rockman can be used during mixdown to generate stereo effects; in this case, you make the multitrack recording directly off the guitar.

Electric Bass

With the electric bass guitar, as always, you first work on the sound of the instrument itself. Substitute new strings if the old ones are dull-sounding. Adjust the intonation and tuning, and adjust the pickup screws (if any) for equal output from each string.

Here are some tips on keeping the bass sound clean and well defined.

- Record the bass direct.
- Use no reverb or echo on the bass.
- If your mixer has sophisticated equalization, cut at 125–400 Hz and/or boost at 1,500–2,000 Hz.
- Have the bass player turn down the bass amp in the studio, leaving it just loud enough to be heard adequately. This reduces muddy-sounding bass leakage into other microphones.
- Better yet, don't use the amp at all. Instead, have the musicians monitor the bass (and each other) with headphones.
- Have the bass player try new strings or a different guitar.
- If it suits the song, the bass player can mute the strings with the side of the hand and play with a pick for extra definition.
- Ask the bass player to use the treble pickup near the bridge.

You might want to try a bass-guitar signal processor such as the Bass Rockman. It has separate three-position switches for equalization, chorus, and sustain, as well as a high-frequency compressor and peak clipper.

Leslie Organ Speaker

This device contains a rotating horn on top for highs and a woofer on the bottom for lows. A typical recording technique is to mike the top and bottom separately, from a few inches to a foot away. Aim the top mic into the louvers. It's often effective to record the rotating horn in stereo with a microphone on either side or with boundary microphones inside the cabinet.

Electric Keyboards and Drum Machines

For maximum clarity, electric pianos, synthesizers, and drum machines are usually recorded direct. If the instrument output is a phone jack, and if your mixer mic input is 3-pin XLR type, use a dircct box between the instrument and your mixer mic input (or use the direct-connection cable shown in Figure 2-6). Note that some inexpensive mixers and recorder-mixers have phone jacks, which can accept a line-level keyboard signal through an ordinary phone-to-phone cable. If using this cable causes hum, try the above mentioned direct-connection cable. If you have enough tracks, record both outputs of stereo keyboards. If the keyboard player has several keyboards plugged into his or her own mixer, you may want to record a premixed signal from that mixer's output.

Drums

Now let's look at recording techniques for the drum set. We'll cover tuning, damping, mic techniques, and equalization.

Tuning and Damping

The first step is to tune the drums so that they sound good without being miked. Try to achieve equal tension on all the tuning lugs. If the drums ring excessively, tape a folded handkerchief or gauze pad near the edge of each drum head. Don't overdo it, though, or the drums will sound like cardboard boxes.

Tune the kick drum fairly loose and use a hard beater for extra click or snap. Remove the front head and put a pillow or blanket inside the drum, pressing against the beater head to tighten the sound. Oil the drum pedal to prevent squeaks. Hold rattling hardware in place with masking tape.

Tune the snare drum with the snares off. A loose top, or batter, head gives a deep, fat sound. A tight batter head sounds bright and crisp. With the bottom, or snare, head loose, the tone is deep with little snare buzz, while a tight snare head yields a crisp snare response. Set the snare tension to the point at which the snare wires begin to "choke" the sound, and then back off a little.

Sometimes a snare drum buzzes in sympathetic vibration with a bass-guitar passage or a tom-tom fill. You may be able to control the buzz by wedging a thick cotton wad or a sock between the snares and the drum stand. Experiment with the position and thickness of the wad for best results.

Miking the Drum Set

Now you're ready to mike the set. For a tight sound, place the mics very close to the edge of each drum head. For a more open, airy sound, move the mics back a few inches, use fewer mics, or mix in some room mics (such as boundary mics or omni condensers) placed several feet away. Sometimes a jazz drum set can be miked adequately with two overhead mics and one kick-drum mic.

Figure 7-3 shows typical microphone placements for a drum set. We'll refer to this figure often in the following text.

Snare

Bring the mic in from the front of the set on a boom. Place it about 1″ above the rim (or 1″ in from the rim), aimed down toward the area the drummer hits (as shown in Figure 7-4).

Figure 7-3. Typical microphone placements for a drum set.

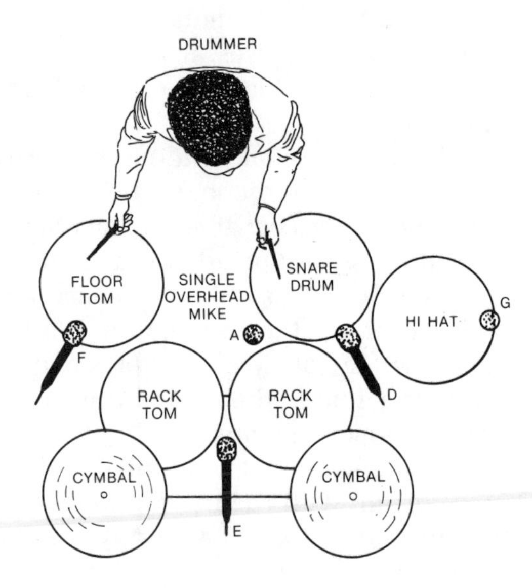

(A) Top view.

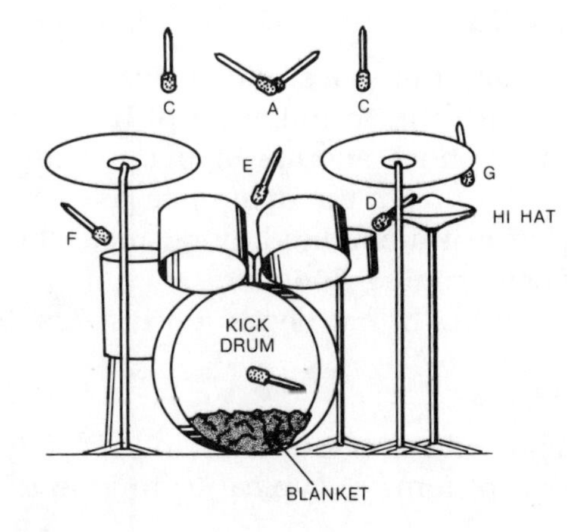

(B) Front view.

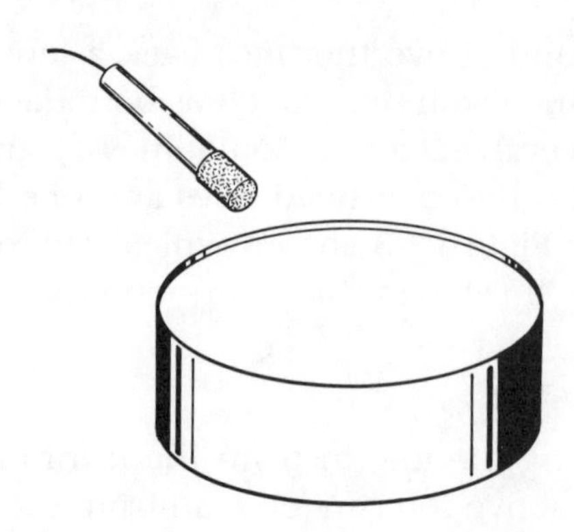

Figure 7-4. Typical mic placement for a snare drum.

You may want to aim the snare mic partly toward the hi hat to pick up both instruments on the one mic, but remember that every time the hi hat closes, it produces a puff of air that can "pop" the snare-drum mic. Place the snare-drum mic so it is not hit by this air puff.

Either a cardioid condenser or cardioid moving-coil microphone works fine—use whichever sounds better for the tune being recorded. Most mics with a cardioid pattern have proximity effect, which boosts bass up close and so adds fullness to the snare beat.

If you want to mike the snare and hi hat separately, bring the boom in under the hi hat, and aim the snare mic away from the hi hat for better isolation. An alternative technique is to attach a miniature condenser mic to the side of the snare drum so that it "looks at" the top head over the rim.

Some engineers like to mike both the top and bottom heads of the snare drum, with the microphones in opposite polarity. A mic under the snare drum gives a "zippy" sound; a mic over the snare drum gives a fuller sound.

Hi Hat

Usually the snare mic or ambience mics pick up enough hi hat, but if you want to mike the hi hat separately, try a cardioid condenser microphone about 6″ above the edge of the hi hat, aiming at the side farthest from the drummer (as shown in Figure 7-5). To avoid the air puff just mentioned, don't mike the hi hat off its edge; mike it from above aiming down.

If the hi hat needs more sizzle, try boosting a little at 10–12 kHz.

Toms

You can mike tom-toms individually or with one mic between each pair of toms. A typical technique is to use a cardioid moving-coil or

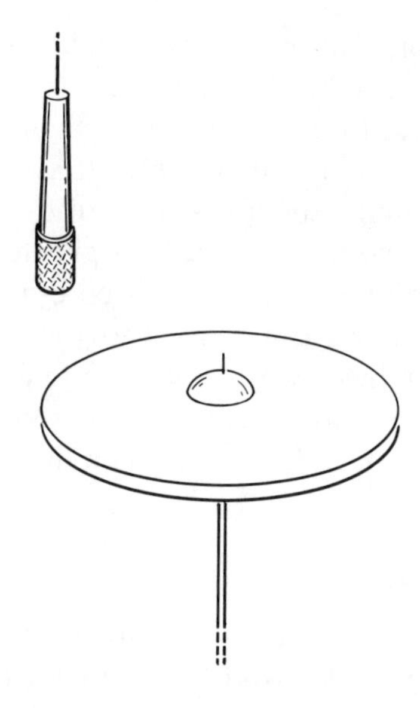

Figure 7-5. Typical mic placement for a hi hat.

Figure 7-6. Typical mic placement for a tom-tom.

condenser mic placed 1″ above the rim (or about 2″ in from the rim), angled down about 45° toward the drum head (as shown in Figure 7-6). Again, the cardioid's up-close bass boost gives a full sound. An alternative setup is to attach mini condenser mics, peeking over the top rim of each tom. You might also try a bidirectional mic between two toms.

Tom-tom mics often pick up too much leakage from the cymbals, which might be heard as a low tone. To reduce cymbal leakage and improve isolation, take the cardioid tom-tom mics and aim their "dead" rear at the cymbals. Note that a supercardioid or hypercardioid mic is partially sensitive to sounds arriving from the rear; it should be placed so that the null of greatest rejection aims at the cymbals.

Another way to reduce cymbal leakage is to remove the bottom heads from the toms and place single mics inside and off-center, a few inches from the batter heads (as shown in Figure 7-7). This also keeps the mics out of the drummer's way. The sound picked up inside the tom-tom has less attack and more ring than the sound picked up outside.

Kick Drum

A cardioid moving-coil mic with an extended low-frequency response is commonly used in the kick drum. Place it inside on a boom, a few inches from where the beater hits (as shown in Figure 7-8). Mic placement close to the beater picks up a hard beater sound; off-center placement picks up more skin tone, and still farther away picks up a boomier shell sound.

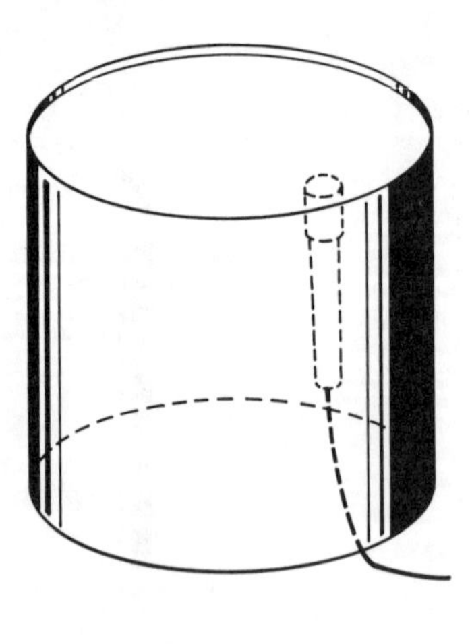

Figure 7-7. Mic placement for a tom-tom that is miked inside.

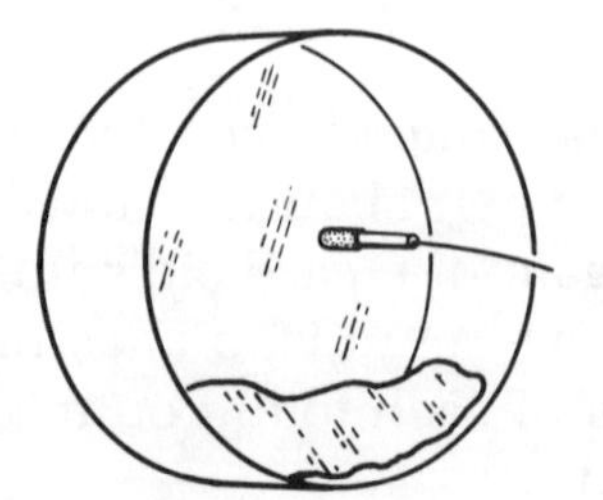

Figure 7-8. Typical mic placement for a kick drum.

You can hang a miniature omnidirectional condenser microphone inside near the beater for a clearly defined attack.

Cutting equalization at around 300–600 Hz helps to remove the "cardboard" sound, and boosting several dB around 2.5–5 kHz adds attack, "click," or "snap."

How should the recorded kick drum sound? They don't call it a *kick* drum for nothing: *thunk!*—a powerful low-end thump plus an attack transient.

Cymbals

Place overhead cardioid condenser mics 1–3′ above the cymbal edges; closer miking picks up a low-frequency ring. Two mics overhead can aim straight down, or they can be angled apart for better isolation, as in Figure 7-9. If your recording will be heard in mono, mount the mic grilles together and angle the mics apart, as in Figure 7-3, position A.

Place the cymbal mics to pick up all the cymbals equally. You won't need to turn up the cymbal mics very high on your mixer, since the cymbals leak into the drum mics.

Recorded cymbals should sound crisp and smooth, not sizzly or harsh.

Recording with Three Microphones

If you're limited in the number of microphones you can spare for the drums, try the setup illustrated in Figure 7-10. It uses two miniature omnidirectional condenser mics and one kick-drum mic. This method works well on small drum sets.

1. Tape or clip one mini mic near the left rack tom and the snare drum. This mic picks up the hi hat, snare, left rack tom, and cymbals.
2. Tape or clip another mini mic near the right rack tom and the floor tom. This mic picks up the right rack tom, floor tom, and cymbals.

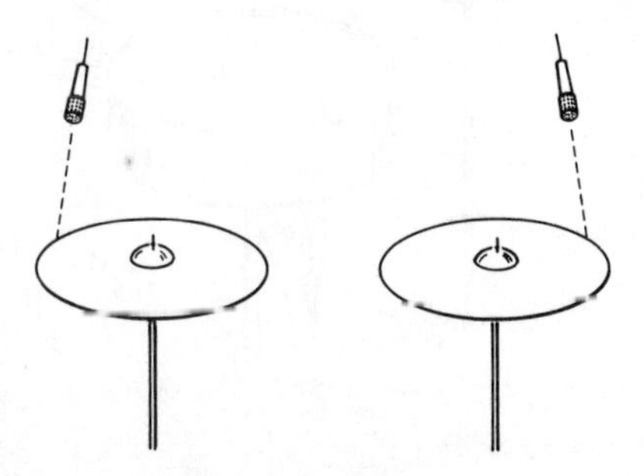

Figure 7-9. Typical cymbal overhead mic placement.

3. Place the third mic inside the kick drum.

With a little bass and treble boost, you'll be surprised at the good sound and even coverage achieved with this simple setup.

Recording with Two Microphones

Here's an even simpler method for mono miking a small set. With some bass and treble boost, the recorded sound can be quite adequate.

Clip a miniature omnidirectional condenser mic to the snare-drum rim about 4″ above the rim, in the center of the set, aimed at the hi hat, as shown in Figure 7-11. Put a second mic in the kick drum.

Equalizing the Drums

Various equalizer settings can enhance the recorded sound of the drums:

- Boost around 200 Hz for fullness on snare drum and rack toms and around 100 Hz on floor toms. Or use a cardioid microphone up close for its bass-boosting proximity effect.
- Roll off some bass on the snare for extra clarity.
- Boost at 5 kHz (or use a mic with a presence peak) on snare and toms to give crispness.
- Boost at 10 kHz or higher on cymbals for brilliance and sizzle, and filter out frequencies below about 500 Hz on cymbals to minimize pickup of low-frequency leakage.

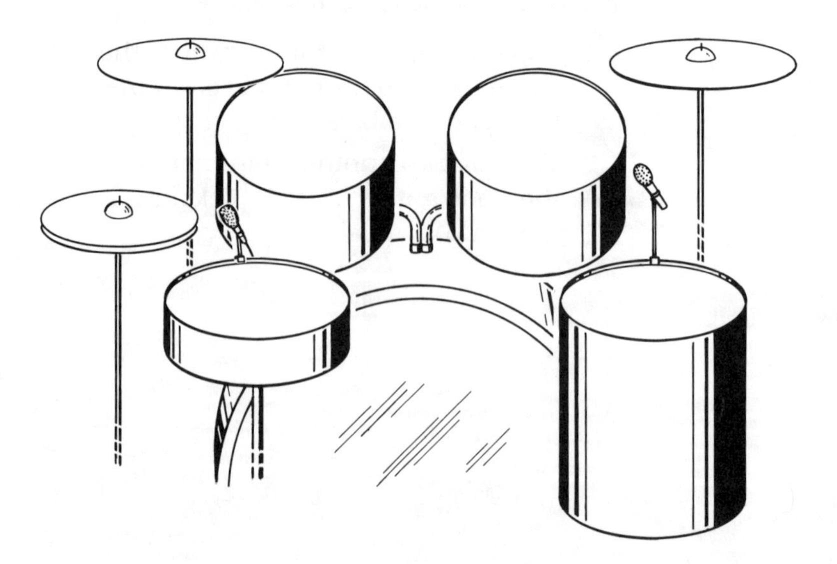

Figure 7-10. Miking a small drum set with three mics (one mic goes in the kick drum).

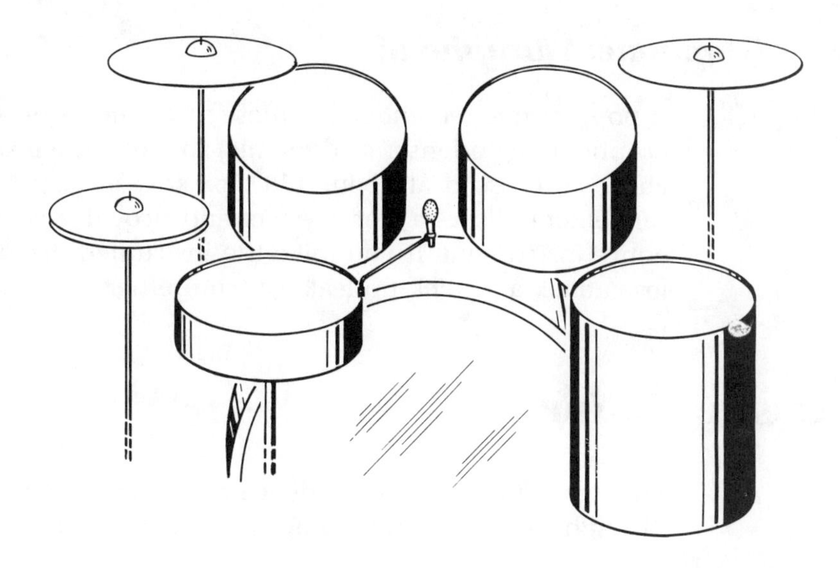

Figure 7-11. Miking a drum set with two microphones (one goes in the kick drum).

- Boost around 2.5 kHz on kick drum for punch, and filter out frequencies above about 5 kHz on bass drum to reduce leakage from cymbals.

Percussion

Percussion instruments in addition to the drum set are a challenge to record accurately. Here are miking techniques for several of these instruments.

Triangle, Tambourine, Guiro, Maracas, Claves

The clarity of a quality condenser microphone makes it a good choice for many percussion instruments. Mike them at least 1′ away to prevent overloading the mic itself, or use a dynamic microphone (moving-coil or ribbon) with an extended high-frequency response.

Congas, Bongos, Timbales

These double drums can be covered with a single microphone between the pair of drums. A cardioid moving-coil microphone with a presence peak gives a full sound with clear attack.

Xylophone, Vibraphone

A popular mic technique for these instruments uses two cardioid microphones (condenser or dynamic) aimed at the instrument about 1½′ above it, crossed at about 135°, or spaced about 2′ apart. This arrangement allows a stereo effect and provides good coverage of the entire instrument. If you mike the instrument from underneath, you lose attack and pick up leakage from other instruments.

Acoustic Guitar

The acoustic guitar has a delicate timbre which we try to capture through careful microphone selection and placement.

Preparation

First prepare the guitar for recording. Use strings designed to reduce finger squeaks, and, for maximum brilliance, replace old strings with new ones. Experiment with different kinds of guitars, picks, and finger picking to achieve a timbre suitable for the song.

Microphone Choice

A condenser microphone with a smooth, extended frequency response from 80 Hz up is often preferred for acoustic guitar. Such a microphone typically gives a clear, detailed quality, in which the plucking of each string is audible within a strummed chord. The reproduced sound usually has all the crispness of the live instrument.

However, the clear pickup of string noise can be distracting in certain songs. You can subdue this fine detail by using a moving-coil microphone, which usually has a slower transient response (the ability to follow sudden changes in acoustic pressure).

Effects of Various Microphone Positions

Miking Near the Sound Hole

If you've ever miked an acoustic guitar close to the sound hole (as in Figure 7-12)—a popular microphone position—you've probably noticed

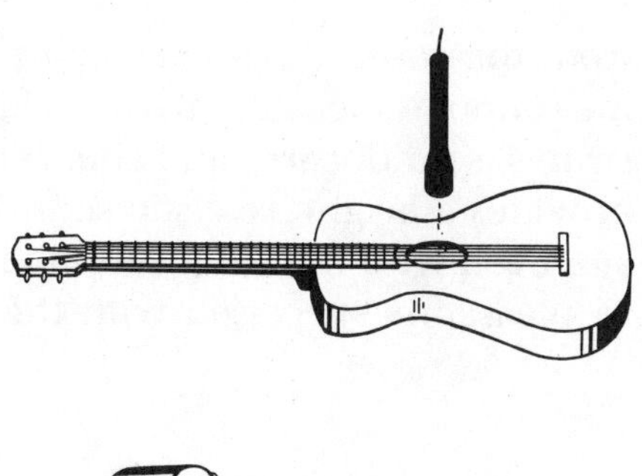

Figure 7-12. Acoustic guitar miked close to the sound hole.

Figure 7-13. Suggested placement of a mini microphone for recording acoustic guitar.

that the recorded guitar doesn't sound much like the real thing. The recording sounds too bassy, boomy, and thumpy. This is mainly because the sound hole and the air inside the guitar resonate at low frequencies (around 80–100 Hz). A microphone placed close to the sound hole (or in it) picks up and emphasizes this resonance, giving a bassy character to the recorded guitar.

Why then is a guitar commonly miked this way? On stage, this position provides maximum loudness before feedback occurs and in the studio, it provides maximum isolation (minimum pickup of other instruments). The acoustic guitar, being a relatively quiet instrument, often requires such a technique to prevent feedback and to reject leakage.

To achieve a more natural sound in this microphone position, roll off the low frequencies on your mixer (say, −10 dB or more at 100 Hz) or use a microphone that has a rolled-off low-frequency response.

Miniature Microphone Placement

A miniature omni condenser microphone taped to the guitar provides good fidelity. A typical mounting position is halfway between the sound hole and bridge, near the low E string (as shown in Figure 7-13). For more isolation, tape a miniature directional mic in the sound hole and, if necessary, roll off the bass on your mixer.

Miking for a Natural Timbre

If leakage is not a problem (as during an overdub), a more natural sound can be achieved by miking the guitar at a distance—say 1½′—

from the sound hole (as shown in Figure 7-14). At this position, the microphone picks up a well-balanced blend of all parts of the guitar: strings, soundboard, and sound hole. A closer placement, which also provides a bright, realistic sound, is 6″ over the top, over the bridge, and even with the front soundboard (shown in Figure 7-15). You may be pleasantly surprised with the sound you get with this technique.

Mandolin, Dobro, Violin

These instruments are constructed somewhat like the acoustic guitar, so many of the microphone techniques for guitar are applicable.

When played, the violin, or fiddle, best radiates high frequencies upward. Consequently, the audience usually hears a duller sound from the violin than the violinist hears. When close-miking the violin, you can avoid the harsh, bright sound the violinist hears by aiming the microphone at the side of the violin. A microphone response down to 200 Hz is sufficient.

Another technique that works well is to clip a miniature microphone holder to the violin's tailpiece and mount the mic a few inches from an f-hole or over the bridge. You can even clip it to the strings on the player's side of the bridge.

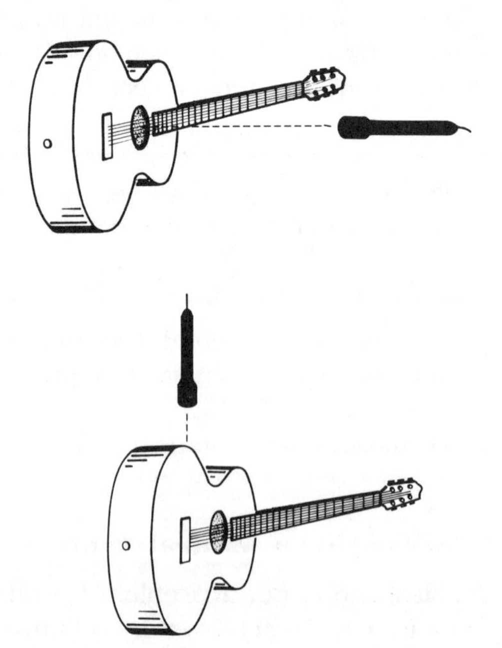

Figure 7-14. An acoustic guitar miked at 1½′ from the sound hole.

Figure 7-15. An acoustic guitar miked 6″ over the top, over the bridge, and even with the front soundboard.

Banjo

The banjo uses a "drum head" to couple the string vibrations to the air. The center of the head vibrates mainly at the head's fundamental frequency, while the harmonics of the head vibration are strongest near the edge. Sometimes the lower notes are reinforced by holes in the flange surrounding the head. To pick up a natural blend of all the parts of the banjo, place a flat-response microphone about 1′ away. Positioning the microphone close to the center of the head produces a rather harsh, thumpy sound (unless you roll off the bass), but provides good isolation. The sound becomes thinner toward the edge of the head.

For maximum isolation, you can tape a miniature omni condenser microphone to the drum head about 1″ in from the bottom edge or clip the microphone holder to the tailpiece and aim the mic at the drum head, about 2″ in from the rim.

Grand Piano

There are many ways to close-mike a piano. Experiment to see what works best for the particular song and instrument. Regardless of technique, always try to get even coverage of all the notes that the pianist plays.

Spaced Microphones

One popular method uses two spaced microphones inside the piano, with the lid on the long stick or removed. One mic goes about 8″ over the treble strings and about 8″ horizontally from the hammers and the other mic goes over the bass strings, about 8″ high and about 2–4′ from the hammers (as shown in Figure 7-16, Position A). These microphones' signals are panned partly toward the left and right for a stereo effect. Alternatively, two boundary microphones can be taped to the underside of the piano lid over the bass and treble strings or taped to the inside of the front edge. Close the lid if necessary for more isolation.

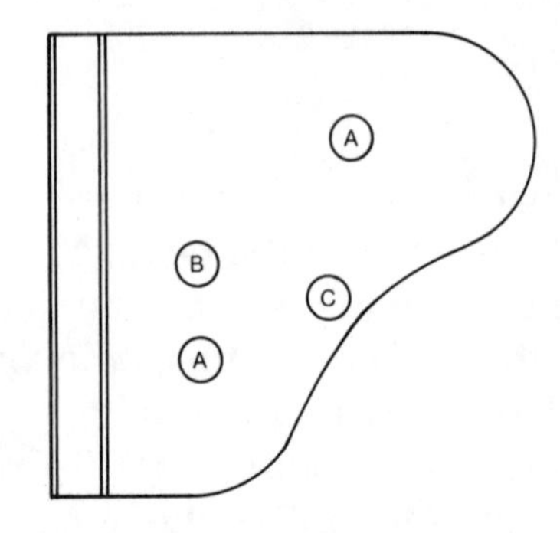

Figure 7-16. Some close-miking positions for a piano.

Coincident Microphones

Spaced microphones can cause phase cancellations when mixed to mono (resulting in a colored tonal balance), so you might want to try coincident miking. Mount a single microphone or a pair of cardioids crossed at 120° and position them about 1′ over the middle of the piano, about 8″ from the hammers (as shown in Figure 7-16, Position B). The closer to the hammers the microphones are placed, the more percussive the attack sounds and the greater the isolation. So, if the sound is too "bangy" and lacks tone, move the mics away from the hammers and toward the tail of the piano.

Again, you might want to tape a boundary microphone to the underside of the piano lid in the middle and close the lid if leakage is excessive.

Sound-Hole Miking

Putting a microphone over a sound hole, with the lid on the long stick, provides good isolation and yields a punchy, constricted sound that can be effective for rock music (shown in Figure 7-16, Position C). Each hole emphasizes the strings closest to it.

For best isolation, aim a microphone into a sound hole, close the lid, and cover the piano with heavy blankets. The tone quality will be unnatural, so you'll have to experiment with equalization if you want a more realistic sound. Special contact pickups for piano are available to further increase isolation.

Bright Sound

You'll often want a bright piano sound. You can improve clarity, sharpness, and attack by boosting frequencies around 5 kHz on your mixer, by using a microphone with a presence peak, or by sticking thumbtacks into the hammer felt. This last method gives a player-piano sound.

Upright Piano

As with a grand piano, each microphone placement for the upright piano produces a different tone quality.

Miking the Panel Area

For a natural sound, remove the panel in front of the player to expose the hammers. Place one mic 8″ from the bass strings and one a like distance from the treble strings. Record in stereo and pan the signals left and right for the desired piano width. If you can spare only one microphone for the piano, be sure to cover the treble strings.

Miking Over the Top

Place two mics just over the open top, one over the bass strings and one over the treble strings. The tone quality is somewhat colored with this technique.

Boundary Miking

Place a boundary mic about 1′ from the soundboard on the floor, either on the player's side or the back side.

Miking the Soundboard

To reduce excessive hammer attack, place a pair of microphones about 8″ from the soundboard, covering the bass and treble sides. The soundboard should be facing into the room, not into a wall.

Miking for Isolation

For extra isolation, place two mics inside the open top. Or tape two mini omni condenser mics to the soundboard and experiment with position for best results. Another alternative is to tape two boundary microphones to the wall 1″ from the soundboard.

Acoustic Bass

The acoustic bass can be recorded in several ways. This instrument produces frequencies as low as 41 Hz, so use a microphone with an extended low-frequency response. For a well-defined sound, place the microphone a few inches out front, above the bridge. Aim it into the treble f-hole for a fuller sound. As always, watch out for proximity effect with a closely placed cardioid microphone.

Here are several techniques to increase isolation and allow the performer freedom of movement.

- Wrap a miniature omni condenser microphone in foam rubber (or in a foam windscreen) and mount it inside an f-hole.
- Wrap a regular microphone in foam padding (except the front grille) and squeeze it behind the bridge or between the tailpiece and the body.
- Try a direct feed from a pickup. This method provides clarity and "bite," but has an electric sound. You can also wrap a condenser lavalier mic in foam and stuff it in an f-hole and mix this microphone with the pickup to round out the tone. You might need to roll off the bass of the f-hole mic.

Brass (Trumpets, Cornets, Trombones, Tubas)

A microphone placed close to, and in front of, the bell picks up a brighter, more "edgy" tone than the audience usually hears. To soften the tone and restore the natural horn sound, try miking the bell at an angle with a flat-response microphone, as shown in Figure 7-17. Or mike it on-axis with a ribbon microphone, which provides a smooth sound. Use a condenser microphone to reproduce a lot of sizzle.

Close microphone placement (about 1′ away) gives a tight sound; distant placement (about 5′ away) yields a fuller, more dramatic sound.

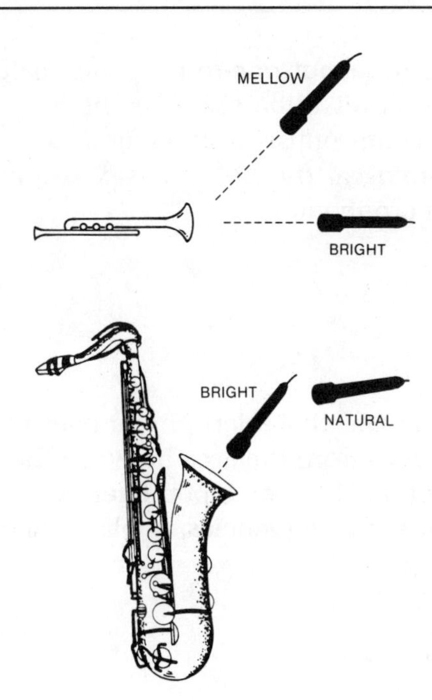

Figure 7-17. Miking for trumpet tone control (top view).

Figure 7-18. Two ways to mike a saxophone.

It's common to mike two or more horns with one microphone. Several players can be grouped around a single omnidirectional mic or around a cardioid mic placed below the group and aimed up. Alternatively, the musicians can play to a boundary microphone taped on a window or on a large panel.

Clarinet, Oboe, English Horn

Most sound radiates not from the bell, but from the holes of these woodwinds. So aim a microphone at the holes from about 1′. Typically, a flat-response dynamic microphone is used.

Saxophone

A sax miked very near the bell (as shown in Figure 7-18) sounds bright, breathy, and rather hard. Mike it there for best isolation. A mic off to the side picks up a quiet sound with poor isolation. For a natural tonal balance, mike the sax from about 1½′ away, aiming at the player's left hand, about one-third to one-half the way down the wind column. Don't mike too close, or the level will vary when the player moves. A compromise position for a close-miking might be just above the bell, aiming at the holes. A sax section can be grouped around a single microphone.

Flute

One effective microphone placement, shown in Figure 7-19, is a few inches from the area between the lips and the first set of finger holes. You might need a pop filter. If you want to reduce breath noises, roll off high frequencies or place the mike farther away.

Harmonica

A popular technique is to place a cardioid moving-coil microphone with a ball grille very close to the harmonica (sometimes held against the instrument by the player). For a bluesy, dirty sound, use a "bullet" type harmonica mic. Or pick up the harmonica with a mic plugged into a guitar amp and record the amp. A condenser mic about 1′ away yields a natural sound.

Harp

Try a cardioid condenser mic aiming at the sound board about 1½′ away. For more isolation, tape a miniature omni condenser mic to the sound board, experimenting with location for the best tonal balance.

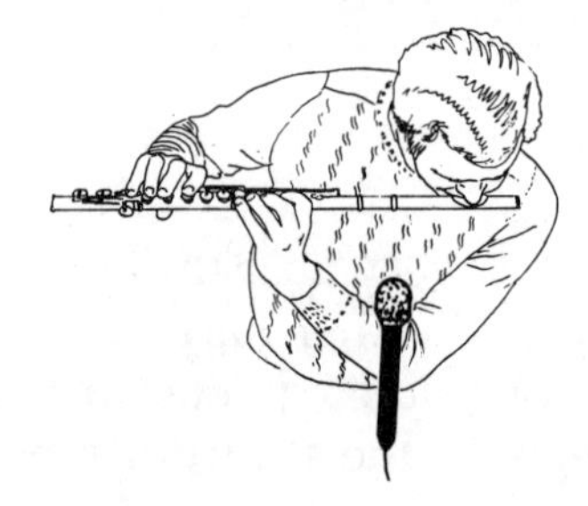

Figure 7-19. An effective microphone arrangement for flute miking.

Vocals

Vocal recording presents a number of problems. Among these are proximity effect, breath pops, wide dynamic range, sibilance, and sound reflections from the lyric sheet. Let's look at these in detail.

Minimizing Proximity Effect

A vocalist on stage has to sing with his or her lips touching the microphone grille in order to reduce feedback. Singing or talking close to a cardioid microphone boosts the low frequencies because of proximity effect. The result is a bassy, boomy tone quality which we've come to accept as a standard sound-reinforced vocal sound. During a recording session, this effect may add robustness to a weak voice; but normally the vocalist should back off at least 8″ from the microphone to restore a natural tone quality. Vocals are typically overdubbed from about 8″ to 2′ with a flat-response condenser microphone, as shown in Figure 7-20.

Close-Miking

If you must record the vocalist simultaneously with the instruments, as in a live recording, you'll probably have to mike him or her very close so that the accompanying musical instruments don't leak into the vocal microphone. A cardioid microphone with a foam pop filter is useful here. To reduce the boominess caused by close placement, roll off the excess bass on your mixer (typically −7 dB at 100 Hz). Some

Figure 7-20. Typical miking technique for a lead vocalist.

microphones have a built-in bass rolloff switch for this purpose. Aiming the microphone up toward the singer's nose will avoid a nasal or "closed-nose" effect.

Minimizing Pop

When a vocalist sings a word with "p" or "t" sounds, an explosive puff of air is forced from the mouth. A microphone placed near the mouth is hit by this air puff and generates an undesirable thump or pop. It can be reduced by placing a foam-plastic pop filter (windscreen) on the microphone. Some microphones have a built-in ball grille screen for pop suppression.

Foam pop filters should be made of special open-cell foam to allow high frequencies to pass through. For this reason, it's better to use a commercially made foam screen than to make one yourself from packing foam, cloth, or socks. Allow a little air space between the foam front and the microphone grille for best pop rejection. Since most pop filters slightly change the frequency response of a microphone, they should not be used for instruments, except for outdoor recording or for dust protection.

Although these devices reduce pop, they do little to minimize breathing sounds or lip noises. Distant miking or some high-frequency rolloff can help with these problems.

An effective way to eliminate popping is to place the microphone well above the singer's mouth level, as shown in Figure 7-20. This way the puffs of air shoot under the microphone. You can also place the microphone off to one side of the mouth as it is in Figure 7-21.

Reducing Wide Dynamic Range

Vocalists often sing too loud or too soft, either blasting the listener or getting buried in the mix. That is, singers generally have a wider

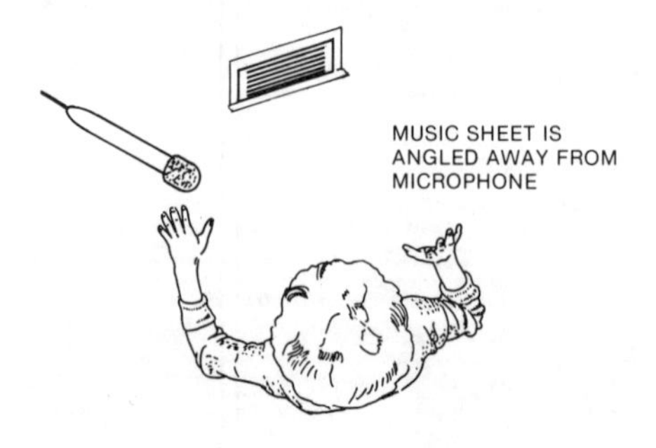
MUSIC SHEET IS
ANGLED AWAY FROM
MICROPHONE

Figure 7-21. Miking a vocalist from the side (top view).

dynamic range than their instrumental backup. To even out these extreme level variations, the vocalist should use proper mic technique: backing away from the microphone on loud notes and coming in closer for soft ones. Or you can ride gain: gently turning the vocalist down on loud notes, and vice versa. The best solution, however, is to pass the vocal signal through a compressor, a device that automatically reduces dynamic range (described in Chapter 6).

A microphone placed close to the vocalist's mouth is sensitive to small changes in miking distance, and loudness will fluctuate if the vocalist fails to keep a constant distance from the microphone or fails to use the mic technique described above. For this reason, it's better to mike the singer from at least 8″. At that distance, small movements of the singer cause less change in loudness. If you must mike close to prevent leakage, have the singer's lips touch the pop filter to maintain a constant distance to the microphone.

Minimizing Sibilance

Sibilance is the emphasis of "s" or "sh" sounds. These are strongest in the 5–10-kHz range. If not controlled, they can easily saturate a tape running at 7½ inches per second (ips) or slower.

To reduce excessive sibilance, use a microphone with a flat response—rather than one with a presence peak—or reduce the highs around 5 kHz on your mixer. A *de-esser* device does this automatically whenever sibilant sounds occur. As an alternative, mike the vocalist from the side rather than in front, as we saw in Figure 7-21, because the sibilants are projected more out front than to the sides.

Reducing Reflections from the Lyric Sheet

Sound reflections from the lyric sheet and music stand can bounce into the microphone along with the direct sound from the vocalist, as is shown in Figure 7-22. The reflections interfere with the direct sound, creating a strange tone quality similar to mild phasing or flanging. To eliminate this effect, place or tape the lyric sheet at the rear of the vocalist's cardioid microphone, perpendicular to the microphone axis (as in Figure 7-20), or mike the vocalist from the side and angle the lyric sheet slightly away from the microphone (as in Figure 7-21). In the first arrangement, reflections entering the rear of the cardioid microphone are rejected. The second method makes reflections bounce away from the microphone.

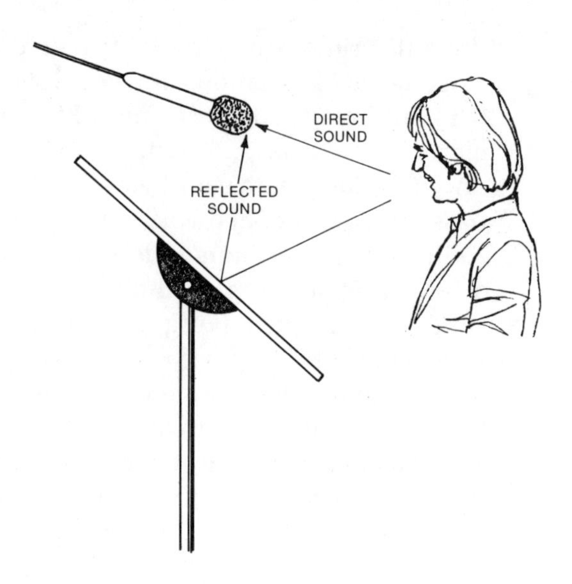

Figure 7-22. Reflections from a music stand can cause interference.

Vocal Effects

Some effects often used on lead vocals are reverberation, echo, and doubling. Although digital reverb is normally used, sometimes you can record reverberation by miking the singer from a distance in a hard-surfaced room. Tape echo (or slap echo on a delay unit) gives a 1950s rock 'n' roll effect. It will sound less mechanical if you roll some highs off the echo signal.

Doubling a vocal provides a fuller sound than a single vocal track. Record a second take of the vocal on an empty track at a slightly different miking distance. During mixdown, mix the second vocal take, at a slightly lower level, with the original. You can also run the vocal signal through a delay device to double it (described in Chapter 6).

Background Vocals

When overdubbing background vocals, you can group two or three singers in front of a microphone. The farther they are from the microphone, the more distant they will sound in the recording. Barbershop or gospel quartets with a good natural blend can be recorded with the stereo setup shown in Figure 7-23 (or with a stereo microphone) from about 2–4′. If their balance is poor, try close-miking them individually, with omnidirectional microphones, and balance them with your mixer.

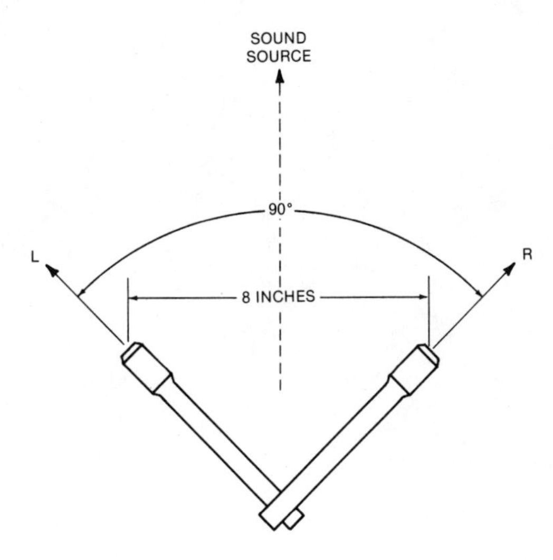

Figure 7-23.
A stereo miking technique (top view).

Summary

We've covered some typical microphone techniques for musical instruments and vocalists. These can serve you as a starting point. Try them out, but invent your own techniques too! If you can capture the power and excitement of amplified instruments and drums and if you can capture the beautiful timbre of acoustic instruments and vocals, you've made a successful recording.

8 Tape Recording

Thanks to the tape recorder, an event as fleeting as a musical performance can be permanently captured and relived. This chapter explains several areas related to tape recording:

- the analog tape recorder: parts, functions, operation, preventive maintenance
- noise-reduction systems
- tape handling, storage, and editing
- the digital tape recorder, R-DAT, and Hi-Fi VCR

The Analog Tape Recorder

An analog tape recorder converts electrical signals into permanent magnetic signals on magnetic tape. The tape itself is a strip of plastic, usually Mylar, with a thin coating of ferric oxide or chromium dioxide particles. These particles have a random magnetic orientation, but they can be aligned into magnetic patterns by the external magnetic field applied during recording. During playback, the tape machine converts the magnetic field on the tape back into an electrical signal.

Recorder Parts and Functions

The tape recorder has three main parts: the heads, the electronics, and the transport.

- The heads are electromagnets that convert electrical signals to magnetic fields, and vice versa.

- The electronics amplify and equalize the signals going to and from the heads.
- The transport pulls the tape past the heads, which contact the tape.

Let's look at each of these parts in detail.

The Heads

Most tape recorders include three heads with differing functions placed left to right: erase, record, and playback (see Figure 8-1). The erase head produces an ultrasonic, oscillating magnetic field. As the tape passes over the erase head, the tape is exposed to a gradually decreasing magnetic field. This orients the magnetic particles randomly and erases any signal on tape.

The record head converts the incoming electrical signal into an analogous varying magnetic field. As the tape passes the record head, the head magnetizes or aligns the tape particles in a pattern that corresponds to the audio signal. This pattern is stored on the tape.

The pattern is a magnetic field. As the tape passes the playback head, the head picks up this magnetic field and converts it back into a corresponding electrical signal. This signal is amplified and sent to speakers, the mixing console, or another tape deck.

Some open-reel recorders, and most cassette decks, use a single head for both recording and playback.

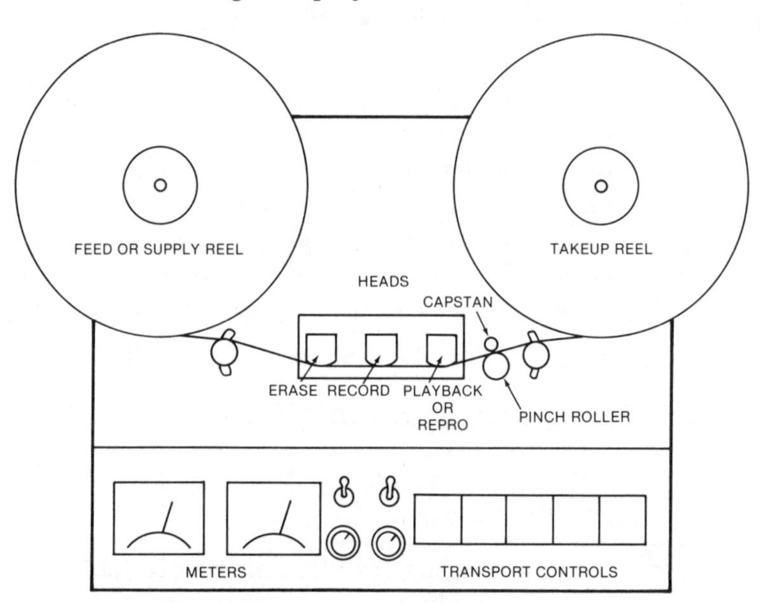

Figure 8-1. Major parts of a typical tape recorder.

There are limits to the signal level that can be recorded. *Tape saturation* occurs when all the magnetic particles are aligned, so that further increases in recording level do not increase the magnetic signal on tape. If the recording level is too low, tape noise (hiss) becomes audible because the recorded signal is weaker in comparison with the random noise signals generated by nonaligned magnetic particles.

It's very important that the heads be correctly aligned with respect to the tape and to each other. The gap in each head (the break in the electromagnet) must be exactly at a right angle to the tape for the best high-frequency response. This is called *azimuth alignment,* which is shown in Figure 8-2. Heads are aligned at the recorder factory and usually stay aligned if you treat the recorder gently.

The Electronics

Tape-recorder electronics perform these functions:

- amplify and equalize the incoming audio signal
- send the audio signal to the record head
- amplify and equalize the signal from the playback head

An ultrasonic oscillator that drives the erase head is among the electronics. The ultrasonic signal, called *bias,* is mixed with the audio fed to the record head to reduce distortion. The amount of bias, which is adjustable, affects the recording's audio level, frequency response, distortion, and *drop-outs* (temporary signal losses).

Bias setting is critical. A setting too high reduces the level recorded on tape and rolls off high frequencies. A setting too low also reduces the level on tape, results in distortion and drop-outs, and raises the high-frequency response. The bias is usually set at the factory for the specific type of tape that you should use with your recorder. If your cassette deck has a bias adjustment, set it according to the manufacturer's instruction booklet.

Figure 8-2. Azimuth alignment.

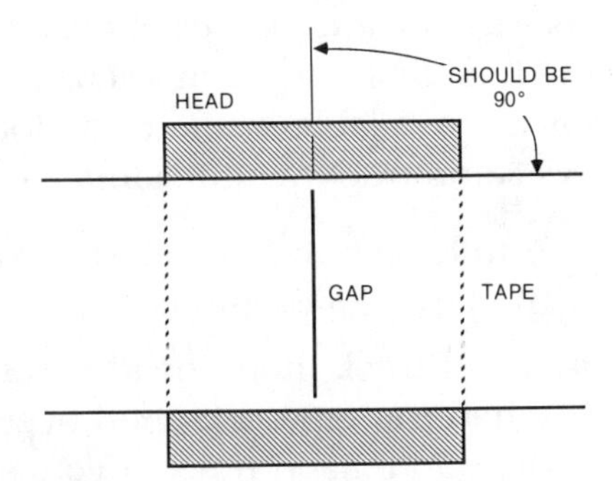

The Transport

The transport moves the tape past the heads. During recording and playback, the transport should move the tape at a constant speed and with constant tape tension. During rewind or fast forward, the tape shuttles rapidly from one reel to the other.

Most professional open-reel machines have three motors in the transport: two for shuttling and tape tension and a third for driving the capstan. The *capstan* is a post that rotates against a *pinch roller.* The tape is pressed between them, and as the capstan rotates, it pulls the tape past the heads. The transport also includes rollers that reduce tape-speed variations (*wow* and *flutter*).

The tape counter usually shows the time elapsed on the tape. A particular point on tape—say, the beginning of a song—can be marked by resetting the tape counter to zero. On some machines, a return-to-zero button shuttles the tape to the zero point and stops automatically. This function is useful for repeated practices of an overdub or a mix.

A professional open-reel tape deck moves tape at 7½, 15, or 30 ips. Cassette decks run at 1⅞ ips or 3¾ ips. As tape speed increases, high-frequency distortion decreases and wow and flutter decrease. A slower tape speed consumes less tape, allowing more running time.

Tracks

A track is a path on tape containing a single channel of audio. The wider the track (that is, the more tape it covers), the greater the signal-to-noise ratio. Doubling the track width improves the signal-to-noise ratio by 3 dB.

Track Width

Recorder heads are available in different configurations. Some can erase, record, and play back over the full width of the tape; some are divided so that they can record two or more independent tracks. Heads are available in the track formats described below. Figure 8-3 shows some track-width standards for ¼″ tape.

- A full-track mono head records over nearly the full width of the tape in one direction.
- A half-track mono head records 1 track in one direction and, when the tape is flipped over, 1 track in the opposite direction. Each track covers approximately one-third of the

tape. The unused third between the tracks is a guard band to prevent crosstalk between tracks.

- A 2-track stereo head records 2 tracks in one direction. This format is used for stereo master tapes. Track width is the same as half-track.
- A quarter-track stereo head records 2 tracks in one direction and, when the tape has been flipped over, 2 tracks in the opposite direction. Consumer stereo cassette decks use this format.
- A multitrack head (4-, 8-, 16-, 24-, or 32-track) records 4 or more tracks in one direction. Recorder-mixers use this format.

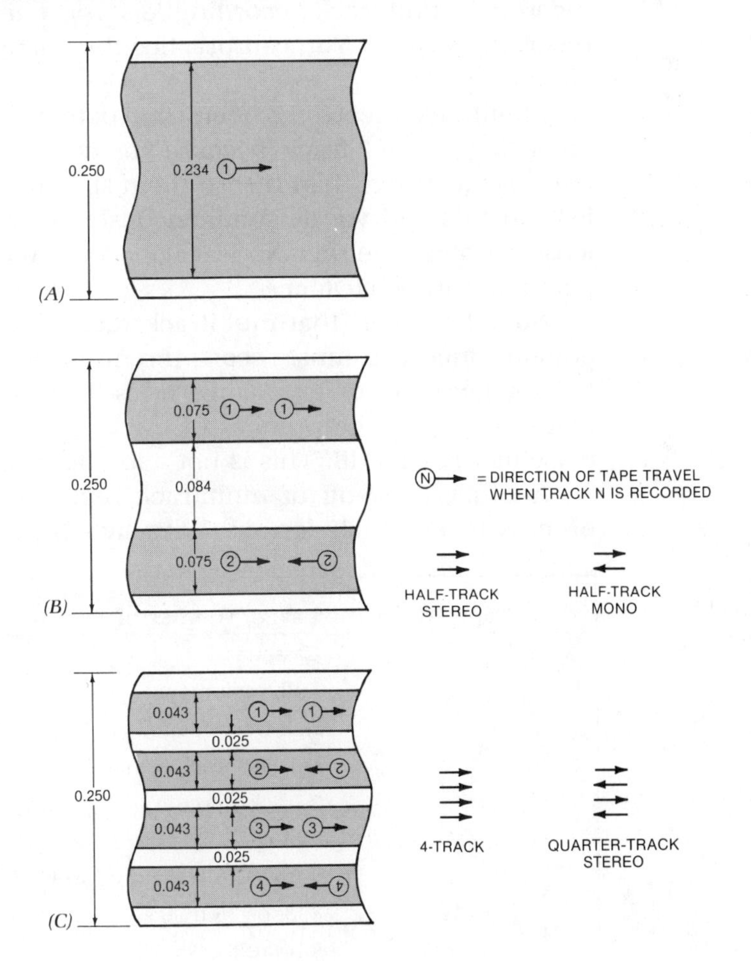

Figure 8-3. Some track-width standards for ¼″ tape.

Tape Widths

Magnetic recording tape comes in various widths to accommodate the track formats, as shown in Table 8-1.

Multitrack and Synchronous Recording

A multitrack machine records 4, 8, 16, 24, or 32 tracks on a single tape. Each track contains the signal of a different instrument or different mix of instruments. The tracks can be recorded all at once, one at a time, or in any combination. After the tracks are recorded, they are combined and balanced through a mixing console. Unlike 2-track recording, multitrack recording lets you fine-tune the mix after the recording session. You can practice the changes in the mix until you get them right.

Multitrack recording offers the potential of clearer sound than recording live to 2-track, because you can overdub instruments without leakage rather than record them all at once. If you record several instruments and vocals simultaneously, leakage can yield a muddy, loose sound in the mix. No leakage occurs when overdubbing, so the final mix can be cleaner.

Note, however, that multitrack recording requires an extra generation, since you must record the multitrack mix on a 2-track tape. Each generation, or tape copy, adds 3 dB of tape hiss and, beyond that, every time the number of tracks used in the mix doubles, the noise increases 3 dB. This is not a lot, but it is audible.

Another trade-off of multitrack recording is that the recording process takes much longer. With live-to-2-track recording, the re-

Table 8-1. Tape Width Options

Tape Width	Usage/Number of Tracks
1/8″	Cassettes
1/4″	Full-track mono
	Half-track mono
	Quarter-track stereo
	2-track stereo
	4-track
	8-track (in some semiprofessional recorders)
1/2″	4, 8, or 16 tracks
1″	8 or 16 tracks
2″	16, 24, or 32 tracks

cording is done when the performance is done. But with multitrack recording, you must record and then do a mixdown. Overdubbing is optional, but is the norm, and adds more time. Plus, multitrack equipment is more complicated and expensive and takes more time to set up. Still, the ability to fine-tune the mix after the session makes multitrack the preferred choice for pop-music recording.

Overdubbing

The tracks can be recorded at different times. To illustrate, suppose that several tracks of music have been recorded on a tape. To overdub, a musician listens to the previously recorded tracks and plays along with them, recording the new part on an unused, or open, track.

If the recorder has three heads—erase, record, and playback—the new part will be delayed relative to the original tracks during playback. The delay occurs because the playback head is a short distance from the record head. While the signal on tape travels from record head to playback head, the travel time delays the monitored sound relative to the part being overdubbed.

To remove this delay and synchronize the original tracks with the overdub, the original tracks are played through the record head. At the same time, the record head records the overdub on an open track. This process is called *simul-sync, selsync,* or *synchronous recording.* Usually you enable it by setting each track's tape-monitor switch to the "sync" position.

Tape decks that combine the record and playback functions in one head do not have a sync problem; previously recorded tracks and overdubs are always synchronized. This is true of most recorders and recorder-mixers for the home market.

Meters and Level Setting

Meters on the tape recorder (one per track) show the record and playback levels. These may be VU meters, VU meters with built-in peak LEDs, or LED bargraph indicators showing peak levels.

The VU Meter

A VU meter is a voltmeter that shows the relative volume or loudness of the audio signal. The meter is calibrated in volume units. When measuring a steady-state sine-wave tone, the volume unit corresponds to the decibel. That is, 1 VU = 1 dB only when a steady tone is applied.

A 0-VU recording level (0 on the record level meter) is the normal operating level of a recorder. It indicates that the optimum recorded flux (magnetic-field strength) is being recorded on tape. Excessive (greater than +3 VU) recording levels saturate the tape, causing distortion. Deficient levels (say, consistently below −10 VU) result in audible tape hiss.

When a complex waveform is applied to a VU meter, the meter reads less than the peak voltage of the waveform. This is because the response of a VU meter is not fast enough to track rapid transients accurately. This inaccuracy can cause problems with level setting. For example, if you record drums at 0 VU on the meter, peaks may be 8 to 14 dB higher, resulting in tape distortion.

So, whenever you record instruments having sharp attacks or a high peak-to-average ratio (such as drums, piano, percussion, or horns), record at −6 to −8 VU to prevent tape distortion. Note that mild distortion on drum peaks (recording "hot") may give a desirable effect. An instrument with a low peak-to-average ratio, such as an organ or flute, can be recorded around +3 VU without audible distortion.

Peak Indicators

Unlike the VU meter, the peak indicator shows peak recording levels more accurately because it responds very rapidly. If your recorder has LED peak indicators, set the levels for all the tracks so that the LEDs only flash occasionally. For setting recording levels, an LED flash takes precedence over the VU meter reading. If the recorder has LED bargraph peak indicators, set all tracks to peak at 0 to +6 dB, depending on the sound source.

Cleaning the Tape Path

Oxide shed by the tape accumulates on the heads. This layer of deposits separates the tape from the heads, causing high-frequency loss and drop-outs. In addition, buildup of oxide on the tape guides, capstan, and pinch roller can cause flutter. So it's essential that you clean the entire tape path frequently: after every 8 hours of use, before alignment, and before every recording session.

Use the cleaning agent recommended by the tape recorder manufacturer. Isopropyl alcohol (from hardware stores or drugstores) and a dense-packed cotton swab are often used. Allow the cleaning fluid to dry before threading tape. Note: some manufacturers recommend using rubber cleaner rather than alcohol for rubber parts to prevent swelling or cracking.

Demagnetizing the Tape Path

Tape heads and tape guides can accumulate residual magnetism, which can partly erase high frequencies, add tape hiss, and cause clicks at splices. You can eliminate this residual magnetism with a tape-head *demagnetizer,* or *degausser,* available from any sound system dealer.

Essentially an electromagnet with a probe tip, the demagnetizer produces a 60-Hz oscillating magnetic field. By touching the probe tip to the heads and tape guides, you magnetize them; by slowly pulling the tip away, you diminish the induced magnetization until no magnetic field is left. Only the gapped demagnetizers are strong enough to be generally effective; the pencil-shaped types may cost less but don't always work as well.

It's a good idea to cover the probe tip with electrical tape to avoid scratching the heads. Following this procedure for using a demagnetizer is critical:

1. Turn off the recorder.
2. When you plug in the demagnetizer, be sure it is at least 3′ from the machine.
3. Bring the demagnetizer slowly to the part to be demagnetized.
4. After touching the part with the probe tip, remove the demagnetizer *slowly* to at least 3′ away so that the induced magnetic field gradually diminishes to zero. Touching the demagnetizer to a head and *quickly* removing it may magnetize the head worse than when you started.
5. Demagnetize each tape head and tape guide one at a time.
6. Turn off the demagnetizer only when it's at least 3′ from the machine.

Demagnetize your machines after every 8 hours of use. The same precautions about slow operation apply to a bulk tape eraser as well.

Alignment

Alignment or *calibration* is the adjustment of tape-recorder circuitry and tape-head azimuth for optimum performance from the particular type of tape being used. It's a complicated procedure not recommended for beginners. Professional recording engineers align their machines periodically to ensure flat frequency response, maximum signal-to-noise ratio, and lowest distortion. Some home and semipro recorders are not designed for easy alignment. The internal pots that need adjustment may not be easily accessible. In that case, the alignment is better left alone, and you use the brand of tape for which the machine was adjusted.

Reducing Print-Through

Print-through is the transfer of a magnetic signal from one layer of tape to the next, causing an echo. If the echo follows the program, it is called *post-echo.* If the echo precedes the program, it is called *pre-echo.* Print-through is especially audible in recordings with many silent passages, such as narration. To minimize print-through:

- Demagnetize the tape path (because stray magnetic fields increase print-through).
- Use 1½-mil tape (thinner tapes increase print-through). Because C-60 cassette tape is thicker than C-90, it is preferred.
- Use noise-reduction devices (discussed later in this chapter).
- Store tapes at temperatures under 80° Fahrenheit, and don't leave tapes on a hot machine (because heat increases print-through).
- Rewind tapes in storage at least once a year. This action allows print-through to decay by separating and realigning adjacent layers of tape.
- Store tapes tail out. That is, after playing or recording a tape, leave it on the take-up reel. Rewinding a tape about 15 minutes before playing helps to reduce print-through that may have occurred during storage. (This measure becomes less effective as the storage time increases.) In addition, tail-out storage results mainly in post-echo, which is less audible than the pre-echo emphasized in tapes that were rewound before being stored.

Operating Precautions

The following operating tips for tape recorders may help you avoid some accidents.

- Don't put the machine in record mode until levels are set. If you record an extremely high-level signal, the crosstalk within the head might erase other tracks.
- Keep tape away from recorder heads when turning the machine on or off or you might put a click on tape.
- Keep degaussers and bulk tape erasers several feet from tapes you don't want to erase.
- Before recording on a track, make sure you won't be erasing something you wanted to keep. Listen to the track first or refer to your track sheet.
- Edge tracks of multitrack tapes are prone to drop-outs due to edge damage. Since drop-outs occur mostly at high frequencies, only use the edge tracks to record an instrument with little high-frequency output (such as bass or kick drum).
- Repeated passes of a recording past the heads may gradually erase high frequencies. You might want to make a copy of the multitrack tape (or a quick two-track mix) with which musicians can practice their overdubs. Go back to the original tape when the musicians are ready to record.
- *Bouncing,* or *ping-ponging* is the process of mixing several tape tracks and recording the mix on an open track. This enables you to erase the original tracks, freeing them up for additional recording. This process tends to lose high-frequency response, so try to limit bounced tracks to bass or midrange instruments.

Noise Reduction

The analog tape recorder adds undesirable tape hiss and print-through to the recorded signal, degrading its clarity. Tape hiss becomes especially audible during a multitrack mixdown because every track mixed in adds to the overall noise level. Noise increases 3 dB whenever the number of tracks in use doubles, assuming they are mixed at equal levels.

Fortunately, noise-reduction devices such as Dolby or dbx—the two major systems—are available to reduce tape hiss and print-through. However, these units do not remove noise in the original signal from the mixing console. If your signal is noisy before you record it, Dolby or dbx will not remove the noise; they work only on tape noise.

One channel of noise reduction is needed per tape track. Noise-reduction units connect between the mixer output buses and the corresponding tape-track inputs and also between the tape-track outputs and the mixer tape inputs, as shown in Figure 8-4. Some open-reel recorders and most cassette recorders have built-in noise reduction permanently connected.

These noise-reduction devices compress the signal during recording and expand it in a complementary fashion during playback. The compressor part of the circuit boosts the recorded level of quiet musical passages. The expander part works in a complementary way during playback, reducing the volume during quiet passages, thereby reducing noise added by the tape. During loud passages (when the noise is masked by the program), the gain returns to normal.

Figure 8-4. Noise reduction applied to multitrack tape and to 2-track master tape.

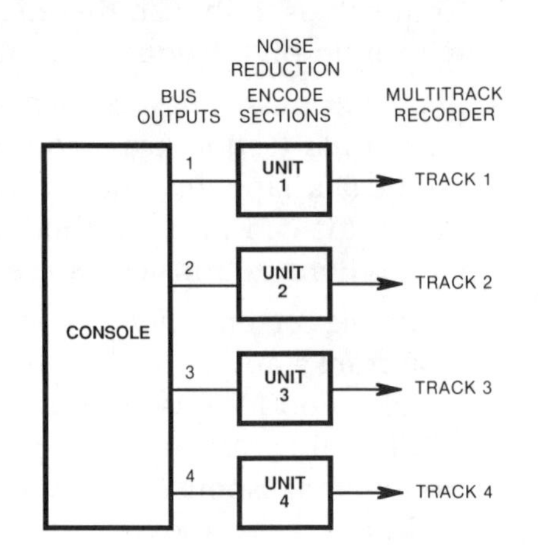

(A) Recording.

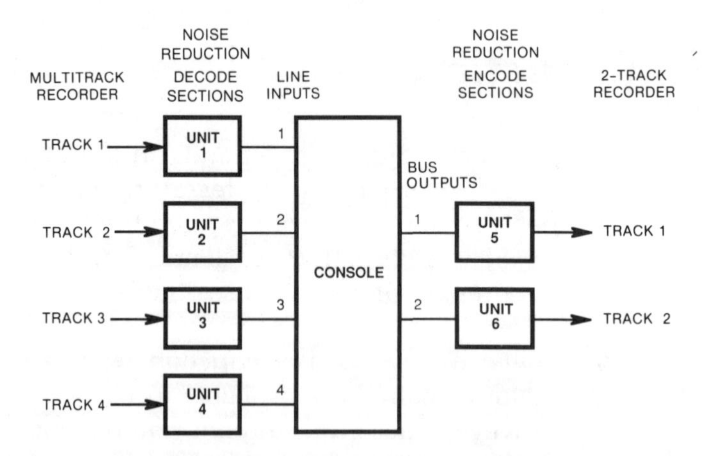

(B) Mixdown.

A compressed tape is described as *encoded;* the expanded tape is called *decoded.* If an encoded tape is played without decoding, the dynamic range and frequency response are altered.

It's important that the encode and decode sections track each other. For example, a 10-dB level change at the input of the encode section should yield a 10-dB level change at the output of the decode section. Otherwise, dynamics will sound unnatural.

dbx

The compression ratio with dbx noise reduction is 2:1. That is, a program with a 90-dB dynamic range is compressed to 45 dB, which is easily handled by a tape recorder with a 60-dB S/N ratio. Then during playback, the dynamic range is expanded to the original 90 dB. Use of dbx improves S/N ratio by 30 dB and increases headroom by 10 dB. The dbx circuit also includes *pre-emphasis* (treble boost) of 12 dB during recording and complementary *de-emphasis* (treble cut) during playback to reduce modulation noise. The dbx system operates at all signal levels and across the entire audible spectrum.

Dolby

Dolby B, found in most cassette decks, operates only at high frequencies to reduce tape hiss by up to 10 dB. Dolby C works over a slightly wider range and reduces noise by up to 20 dB. Dolby A and Dolby SR are used in professional studios.

Dolby vs. dbx

Dolby and dbx each have advantages and disadvantages. Compared to Dolby, dbx provides more noise reduction. On the other hand, dbx exaggerates drop-outs more than Dolby does. Dolbyized recordings are relatively free of noise "breathing," which is sometimes audible on a dbx-encoded tape as fuzziness accompanying bass or percussion tracks.

Using Noise Reduction

Dolby- and dbx-encoded tapes are not compatible with each other and cannot be played properly without decoding through the appropriate unit. So when you send your demo cassette tape to someone else, use Dolby B because all consumer stereo cassette decks have Dolby B. You can use whichever system you want on your multitrack tape.

When you copy a tape recorded with noise reduction switch in the noise reduction on the playback deck and also on the recording deck.

Matching the Mixer Meters and Recorder Meters

In a recorder-mixer, the mixer and recorder meters are the same. But if you're using a mixer separate from a recorder, it's common practice to set the mixer meters and recorder meters to match. That way you only have to watch the mixer meters while recording. Also, when the mixer and recorder are both peaking around 0 VU, this condition prevents excessive noise and distortion in both units.

Ideally, meter matching is done with a steady tone from a signal generator or a synthesizer note (C or B two octaves above middle C). Otherwise, hum a steady tone into a microphone plugged into your mixer, and set the levels on the mixer and tape deck to 0.

If your tape-deck meters move at a different pace from your mixer meters, you'll have to watch the tape-deck meters while recording.

Also, if you're using dbx noise reduction, matching the meters can be confusing because the encoded signal from the dbx is compressed. The recorder meters will wiggle less than the mixer meters. Follow the calibration instructions in the dbx manual and watch the recorder meters while recording.

Once the mixer and recorder are calibrated to match each other at 0 VU, leave the recorder controls alone and set your levels with the mixer faders only.

Tape Handling and Storage

Careful handling and storage of tape reels is essential to avoid damaging the tape and the signals recorded on it.

If you examine a reel of used recording tape, you may see some edges or layers of tape sticking out of the tape pack. These edges can be crushed by pressure from the reel flanges, causing drop-outs and high-frequency loss. For this reason, never hold a reel of tape by squeezing the flanges together. Instead, hold the reel in one hand by putting your fingers in the hub and your thumb on the flange edges, or hold the reel in two hands with fingers on the flange edges, as shown in Figure 8-5.

To prevent edge damage during storage, leave tapes tail-out to ensure a smooth tape pack. Repair or discard any reel with a bent flange. Reels left in the open can collect dust, so keep them in boxes and store the boxes vertically—do not stack them. The preferred storage conditions are 60–75° Fahrenheit, with 35–50% relative humidity. Keep tapes away from magnetic fields such as those caused by speakers, headphones, or telephones.

Editing and Leadering

Editing is the cutting and rejoining of open-reel magnetic tape to delete unwanted material, to rearrange material into the desired sequence, or to insert leader tape. If you're doing all your work on cassettes, you can skip this section.

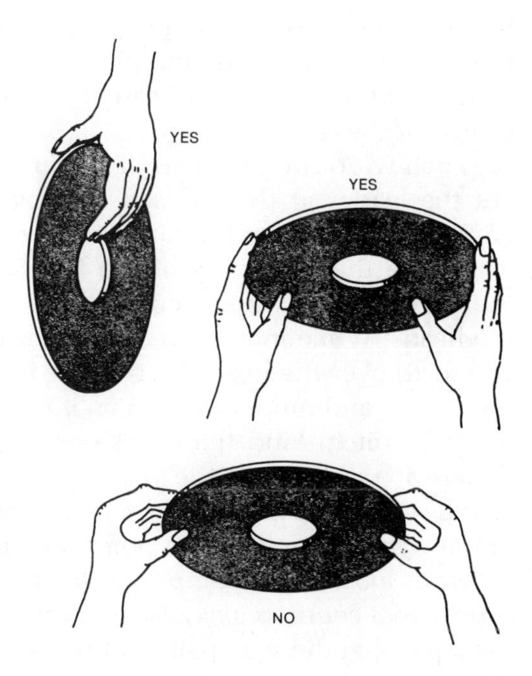

Figure 8-5. Handling tape reels.

Equipment and Preparation

Editing requires the following materials: demagnetized single-edge razor blades, a light-colored grease pencil, splicing tape, leader tape, and an editing block.

Leader tape is plastic or paper tape without an oxide coating. It is used as a spacer between takes (yielding silence between recorded songs). Plastic leader is preferred over paper because paper can absorb humidity and become warped during long storage. An *editing block* holds the tape during the splicing operation. It's easier to use than a tape splicer (which has hold-down tabs) and allows more precise cuts.

To avoid getting oily spots on the tape, wash your hands before editing. Cut several 1″ pieces of splicing tape and stick them on the edge of the tape deck or table. Also cut several sections of leader at the 45° slot in the editing block. A typical leader length between songs is 4 seconds, which is 60″ long for 15 ips or 30″ long for 7½ ips. While editing, try to hold the magnetic tape lightly by the edges.

Leadering

Suppose you've recorded a reel full of takes, and you want to remove the outtakes, count-offs, and noises between the good takes. You also want to insert leader between each song. This process, called *leadering,* can be done as follows.

First, wind several turns of leader onto an empty takeup reel and cut the leader at the 45° slot. Remove this takeup reel, put on an empty one, and play the tape to be edited.

Locate the beginning of the first song's best take. Stop the tape there. Put the machine in cue or edit mode so the tape presses against the heads. While monitoring the recorder output, rock the tape back and forth over the heads by rotating both reels by hand; first rapidly, then more and more slowly. You'll hear the music slowed down and lower in pitch. Find the exact point on tape where the song starts (where it passes over the playback-head gap). Align the beginning sound with the gap. Using the grease pencil, mark the tape about ½″ to the right of the gap (at a point on tape just before the song starts).

Next, loosen or "dump" the tape by simultaneously rotating the supply reel counterclockwise and the takeup reel clockwise. Remove the tape from the tape path and press it into the splicing block, oxide side down. Align the mark with the 45°-angled slot, as shown in Figure 8-6A. Don't use the 90° slot because such an abrupt cut can cause a

Figure 8-6. Aligning edit marks with cutting slot.

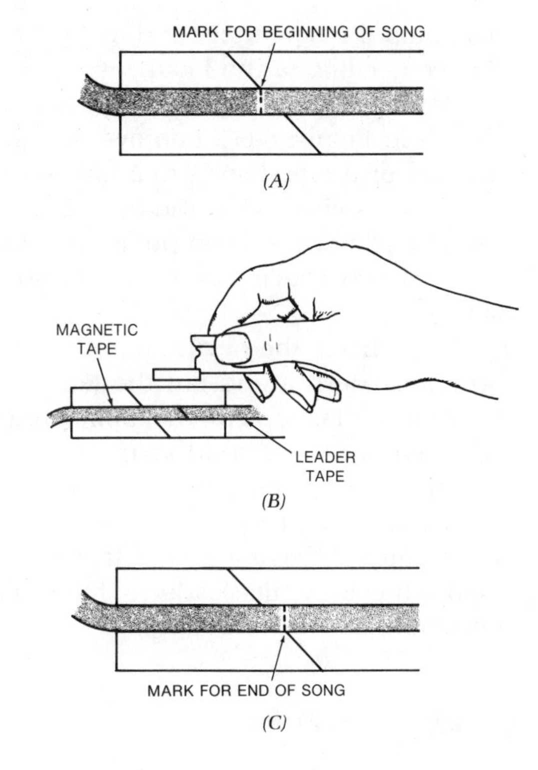

pop noise at the splice. Slice through the tape with a razor blade, drawing the blade toward you.

Remove the unwanted tape to the right of the cut and put the takeup reel aside. Slide the cut end of the tape to the right of the editing-block slot, as in Figure 8-6B. Put on the takeup reel containing the turns of leader tape, and insert the end of the leader into the right half of the block. Slide together the ends of the leader tape and magnetic tape so that they butt or touch together with no overlap.

Now, take a piece of splicing tape and stick a corner of it onto a hand-held razor blade. Align the splicing-tape piece parallel to the recording tape, as shown in Figure 8-6B. Apply the piece over the cut onto the nonoxide side and adhere it by rubbing with a fingernail.

Slide the splice out of the block by gently popping the tape out of the block. Pull up on the ends of the tape extending from both sides of the block, twist the tape toward you while pulling, and check that there is no gap or overlap at the splice.

Now wind the tape onto the takeup reel and locate the ending of the first song. As it ends, turn up the monitors in order to listen for the point where the reverberant "tail" of the music fades into tape hiss. Stop the tape there and mark it lightly at the playback-head gap (the center line of the head).

After pressing the tape into the block, cut the tape at the mark, as shown in Figure 8-6C. Remove the tape to the left of the cut. Splice the end of the first song to a four-second length of leader and again check the splice. Wind the first song and the leader onto the takeup reel and remove it. Then put on the takeup reel containing unwanted material you previously set aside. Splice it to the rest of the master tape.

Next, locate the beginning of the next good take you want in the program. Mark it and cut the tape. Put the reel containing the first song on the takeup spindle. Splice the tail end of the leader onto the beginning of the second song, then wind the second song onto the takeup reel. You now have two songs joined by leader tape.

Repeat this process until all the good takes are joined by leader. Then you will have a reel of tape with several songs separated from each other by white leader, which makes it easy to find the desired selection.

Joining Different Takes

What if you want to join the verse of Take 1 to the chorus of Take 2? You'll have to cut into both takes at the same point in the song, then join them. It takes practice to make an inaudible splice in this manner, but it's done every day in professional studios.

The two takes must match in tempo, balance, and level for the edit to be undetectable. To mask any clicks occurring at the splice, cut the tape just before a beat—say, at the beginning of a drum attack—or cut into a pause. If you cut into a continuous sound such as a steady chord, a cymbal ring, or reverberation, the splice will be noticeable.

Let's run through the procedure. First play Take 1 and locate the point where you want it to stop and Take 2 to start—say, at the beginning of the chorus. Stop the tape there. Put the recorder in cue or edit mode, rock the tape, and try to identify a beat or attack transient. At the point on tape where this beat just starts to cross the playback-head gap, mark the tape. Cut the tape at the mark and remove the takeup reel containing the verse of Take 1.

Next, put on an empty takeup reel, thread the master tape, and fast-wind to Take 2. Find the same spot in Take 2 that you marked in Take 1. Mark and cut it. Using splicing tape, join Take 2 (in the supply reel) to Take 1 (in the set aside takeup reel). Again, check that there is no gap and no overlap at the splice.

Play the spliced area to see if the edit is detectable. If not, congratulations! It should sound like a single take. If Take 2 comes in a little late, carefully remove the splice and cut out just a little tape surrounding the cut. Resplice and listen again.

More Editing Tips

Suppose you've recorded most of a good take, but at some point the musicians make a mistake and stop playing. Rather than repeating the entire song, the musicians can start playing a little before the point where they stopped and then finish the song. You can splice the two segments into one complete and perfect take. Editing is also useful for inserting sound effects in the middle of a song or for making tape loops. You can even record a difficult mixdown in segments, then edit the segments together.

Digital Recording, DAT, and Hi-Fi VCR

The type of recorder that we've been talking about is an analog recorder. That is, the magnetic particles on tape are oriented in patterns corresponding to the audio waveform. The drawbacks of this system are tape hiss, tape distortion, frequency-response errors, and speed variations (wow and flutter). Recently, digital recorders have been developed that eliminate these problems. The digital recorder is beyond the scope of this book, but we'll look at a brief overview.

A digital recorder measures (samples) the voltage of the incoming waveform 48,000 times a second. Numbers are assigned to these voltages. This process is called *analog-to-digital (A/D) conversion.* The machine records the numbers on tape in binary code (1s and 0s), which is accurate to 16 digits (bits). The binary numbers are in the form of a modulated square wave recorded at maximum level.

During playback, the machine reads the binary numbers and converts them to the original sampled voltages—a process called *digital-to-analog (D/A) conversion.* Finally, the varying voltage levels are smoothed back into the original audio waveform by a lowpass filter.

Since the digital playback head reads only two binary numbers, it is insensitive to tape hiss and distortion. Numbers are read into a buffer memory and read out at a constant rate, eliminating speed variations. The resulting freedom from noise, distortion, print-through, wow, and flutter makes digital recordings sound extremely clean and clear. Unlike analog recordings, digital recordings can be copied with little or no degradation in quality. Lost data is restored by error-correction circuitry.

Currently, digital recorders are quite expensive. For stereo mastering, there's an alternative to a digital recorder: a *digital-audio processor* with pulse code modulation (such as shown in Figure 8-7). This device digitally encodes the audio and records it on a standard video cassette tape. You connect the analog signal to be recorded to the processor inputs and connect the processor outputs (a modulated radio-frequency signal) to the inputs of any video cassette recorder. Then you set levels and start recording. During playback, the modulated video signal feeds the processor and analog audio comes out.

Another alternative for 2-track mastering is available: the R-DAT (Rotating head Digital Audio Tape) cassette recorder (now called DAT). It records a digital signal on a special size cassette tape, and the resulting sound quality is on a par with the compact disc. Up to 2.2 hours of recording time is possible on a single cassette, and access time is very fast. Editing on DAT is difficult or impossible however. Rather than putting leader between songs, you allow a few seconds of blank tape between recordings of mixdowns.

Yet another alternative is a VHS Hi-Fi or Beta Hi-Fi VCR (videocassette recorder). The audio signal is recorded as an ultrasonic tone that is frequency modulated—the same principle used in FM radio broadcasting. The sound quality is nearly as good as that of a digital re-

Figure 8-7.
A digital-audio processor, the Sony PCM-F1.

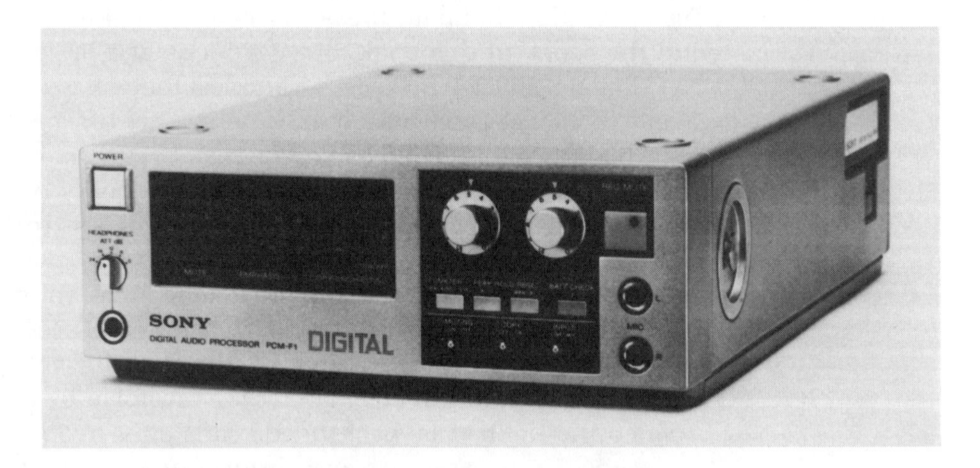

corder, yet the VCR costs much less and allows up to 6 hours of recording time per cassette.

Thanks to digital recording and the Hi-Fi VCR, the original goal of tape recording has finally been achieved: to accurately store and reproduce our audio creations.

9 Session Procedures

By now you've collected your recording equipment, connected it, learned about recorder-mixer features, signal processors, microphone techniques, and tape recorders. Now you're ready to make a multi-track recording. This chapter covers typical operating procedures for a home recording session using a cassette recorder-mixer, but most of the principles also apply to an open-reel multitrack recorder and separate mixer.

There are three stages in making a multitrack recording: recording, overdubbing, and mixdown. First you record the rhythm instruments—drums, bass, guitar, keyboards. Then you add vocals and other instruments during the overdubbing stage. Finally you mix all four recorded tracks to 2-track stereo in the mixdown stage. Record this stereo mix on an external 2-track machine. If you record the mix on an open-reel machine, add blank leader tape before and after the program and between songs.

Presession Planning

Before the session starts, you'll plan track assignments, overdubs, and microphone selection.

Planning the Recording Schedule

The first step is to make a list of the instruments and vocals that will be used in each song. Include details such as the number of tom-toms, whether acoustic or electric guitars will be used, and so on.

Next, decide which instruments to record at the same time and which to overdub one at a time. It's common to record the instruments in the following order, but there are always exceptions:

1. Loud rhythm instruments—bass, drums, electric guitars, keyboards. The lead vocalist usually sings a *reference vocal* or *scratch vocal* along with the rhythm section so the musicians can get a feel for the tune and keep track of where they are in the song. Whether to record the vocalist is up to you.
2. Quiet rhythm instruments—acoustic guitar and piano.
3. Lead vocal, doubled lead vocal (if desired).
4. Backup vocals (in stereo).
5. Other overdubs—solos, percussion, synthesizer, sound effects.

It's best to record loud instruments and quiet instruments separately. If you record a drum set and an acoustic guitar at the same time, for example, the drums will sound distant and muddy because the guitar mic picks up the drums at a distance.

Make a list of your planned sequence of recording basic tracks and overdubs. An example is shown in Figure 9-1.

Track Assignments

Once the instrumentation and the order of recording are clear, you can plan your track assignments. Decide which instruments will go on which tracks of the multitrack recorder, and write this information on a *track sheet,* like that shown in Figure 9-2. Note that the outer tracks are most prone to drop-outs at high frequencies and so are usually reserved for bass and kick drum.

If you have more instruments than tracks, you'll have to decide what groups of instruments to put on each track. In a 4-track recording, for example, you might record a stereo mix of the rhythm section on tracks 1 and 2, then overdub vocals and solos on tracks 3 and 4. Or you might put guitars on track 1, bass and drums on track 2, vocals on track 3, and keyboards on track 4.

Remember that each track is mono. When several instruments are assigned to the same track, you can't separate their images in the stereo stage. That is, you can't pan them to different positions; all the instruments on one track will sound like they're occupying the same

Figure 9-1. A schedule for a multitrack recording session.

RECORDING SCHEDULE

1. Song Title: "Mr. Potato Head."
   Instrumentation: Bass, drums, electric rhythm guitar, electric lead guitar, acoustic piano, sax, lead vocal.
   Comments: Record rhythm section together with reference vocal. Overdub sax, acoustic piano, lead vocal later.
2. Song Title: "Sambatina"
   Instrumentation: Bass, drums, acoustic guitar, percussion, synthesizer.
   Comments: Record rhythm section with scratch acoustic guitar. Overdub acoustic guitar, percussion, synthesizer.
3. Song Title: "Mr. Potato Head."
   Overdubs: (1) Acoustic piano, (2) lead vocal, (3) sax.
4. Song Title: "Sambatina."
   Overdubs: (1) Acoustic guitar, (2) synthesizer, (3) percussion.
5. Mix: "Mr. Potato Head."
   Comments: Add 80 ms delay to toms.
   Double lead guitar in stereo.
   Increase reverb on sax during solo.
6. Mix: "Sambatina."
   Comments: Add flanger to bass on intro only.
   Manually flange percussion.

Figure 9-2. A track sheet.

TRACK SHEET

Song Title: "Dig up Nebraska."

Track 1: bass
2: rhythm guitar
3: lead guitar
4: piano
5: lead vocal
6: drums L
7: drums R
8: kick

Song Title: "Sidewalk Blues (Experimental)."

Track 1: banjo-bass
2: vocal
3: doubled vocal
4: exponential horns
5: tinker toys (percussion)
6: slinky (percussion)
7: foot stomps
8: spare

point in space. For this reason, you may want to do a stereo mix of the rhythm section on, say, tracks 1 and 2 and then overdub vocals and solos on tracks 3 and 4. If you want a stereo drum mix, it must be recorded on two tracks. A mono guitar track can be made stereo during mixdown by running it through a stereo chorus or stereo synthesizing device.

It's possible to overdub more than four parts on a 4-track recorder by bouncing, as we saw in Chapter 8. To do this, you mix several tracks and record them on an unused track. Then you erase the original tracks while recording new instruments over them. Be sure to leave a track open (unrecorded) if you plan to bounce tracks. Detailed procedures for this are given later in the chapter.

Microphone Input List

Now make up a microphone input list similar to that shown in Table 9-1. Write a column of numbers on the left corresponding to each numbered mixer input. Next to each input number write the name of the instrument assigned to that input. Finally, write next to each instrument the microphone(s) or direct box you plan to use on that instrument.

Be flexible in your microphone choices—you may need to experiment with various microphones during the session to find one giving the best sound with the least equalization.

Setup

Let's run though a typical recording procedure for a small rock band. **Important:** First clean the tape heads and rubber pinch roller with

Table 9-1. Microphone Input List

Input	Instrument	Microphone
1	Bass	Direct
2	Kick	EV RE-20
3	Snare and Hi Hat	AKG C451
4	Drums Overhead L	Shure SM81
5	Drums Overhead R	Shure SM81
6	Rack Toms	Sennheiser MD421
7	Floor Toms	Sennheiser MD421
8	Electric Lead Guitar	Shure SM57
9	Electric Lead Guitar	Direct
10	Piano	Crown PZM-30R
11	Reference Vocal	Beyer M500
12	Spare	—

a cotton swab and the cleaner recommended by the manufacturer (usually isopropyl alcohol, available at drugstores or hardware stores). Allow the heads to dry. Insert a high-bias (chrome) cassette tape of C-60 length.

To prevent hum, power the electric instruments and your recording equipment from the same outlet strip. First be sure that the sum of the equipment fuse ratings doesn't exceed the amperage rating for that circuit.

Next, set up the musical instruments. Listen to their live sound and do what you can to improve it. A dull-sounding guitar may need new strings, a noisy guitar amp may need new tubes, and so on. If the room is too "live" or reverberant, add acoustical absorbers such as those described at the end of Chapter 2.

Place microphones and plug them into the mic inputs. If you record more than one instrument at a time, you'll need to mike very close to avoid leakage from other instruments. Leakage makes a recording sound distant and muddy, but it can be avoided by

- close miking
- using directional microphones (such as cardioids)
- recording direct
- recording in an acoustically dead room with acoustic absorbers

Connect a cable between each synthesizer-drum machine output and a line input on the recorder-mixer. If this connection causes hum, use a direct box or the direct-connection cable shown in Figure 2-6. Some keyboardists run all their instruments into a mixer; you can connect to their keyboard mixer output.

Using a strip of masking tape, label the input faders of your recording mixer according to the instruments they affect. Set the input selectors to "mic" or "line" depending on what is plugged into each input. If there are no input selectors, turn down the "trim" control for line-level signals.

Set all the other mixer controls to "off," "flat," or "zero" to establish a point of reference and to avoid surprises later.

Plug in headphones (or use monitor speakers) to hear what you're recording. Turn up the headphone or monitor volume control. Set the "monitor select" switch in the position that lets you hear the live signals you're recording.

Assign each instrument to the track you've noted on your track sheet. Label each track's meter according to what is being recorded on that track.

You may want to turn up the treble (high-frequency EQ) a little on all instruments (except bass and kick drum) to compensate for losses during recording and to improve the S/N ratio.

Set the master fader(s) about ¾ up (at 0, or at the shaded portion of fader travel). This position is called *design center.* Do the same for the input faders in use. Play the instruments one at a time as loud as they're going to be played, and gradually turn down the "trim" controls until the input-overload (clip or peak) lights just stop flashing. Or set them so that the meters peak around 0. If your recorder-mixer has no "trim" control, simply proceed to the next step.

Using the input faders, set the recording levels to peak around 0 (or, for percussive instruments, around −6 to −8 VU). If the VU meters contain peak LEDs, set the recording levels so that the peak LEDs flash only occasionally. If your tape deck has LED bargraph meters, peak at about +3 dB for all instruments.

If you have several microphones assigned to one bus, you'll need to set up a *submix* of those microphone signals. Monitor that bus and turn up the faders for those microphones. As the musicians play, set the desired balance between microphones with the faders while maintaining a proper recording level. Your mixer might have a submaster, bus master, or group fader for each output bus that controls the overall level of the submix.

Next, adjust the monitor mix for the desired balance among tracks as you hear them over headphones or loudspeakers. That is, adjust the "gain" and "pan" controls as desired in the monitor mixer (which might be called "monmix" or "tape cue"). In some inexpensive units, you cannot do a monitor mix of live microphone signals. In other units, the monitor mix is set with the "output" knobs in each input module. The monitor mix does not affect the recording levels; it just affects what you're listening to so that you can simulate the final mix while recording.

Set the tracks you want to record to "record ready" mode, and set the tape counter to 000.

Recording

Start the tape in record mode. Announce on tape the name of the tune and the take number. Have the keyboardist play the key-note of the song (for tuning other instruments later). Then the drummer counts off the beat, and the musicians start playing. Do a trial recording to check again for input overload and excessive recording levels.

The lead vocalist might sing along with the instruments so that the musicians can keep their place in the music and get a feel for the song. The singer is picked up with a mic and monitored by the other musicians over headphones. The scratch vocal is usually not recorded; vocals will be overdubbed later.

At the end of the song, the musicians should be silent for several seconds after the last note. Or, if the song will end in a fade-out, the musicians should continue playing for about a minute so that there will be enough material for a fade-out during mixdown.

When the song is done, rewind the tape using the return-to-zero function. Set the input selectors to "tape," "track," or "remix," and play back the recording. You can make a rough mix by using the "monmix" (or "tape cue") controls. Don't expect this playback to sound like the finished product—you'll refine the sound during mixdown.

If necessary, record several performances or takes and pick the best one. Keep track of the takes on a *take sheet.* Note the name of the tune, the take number, and whether the take was complete. You'll find a code useful to indicate whether the take was a false start, nearly completed, a "keeper," and so on. Figure 9-3 is a sample take sheet.

Vocal Overdubs

Do your overdubs after the rhythm tracks for all the songs are recorded. The lead vocalist listens to the previously recorded tracks over

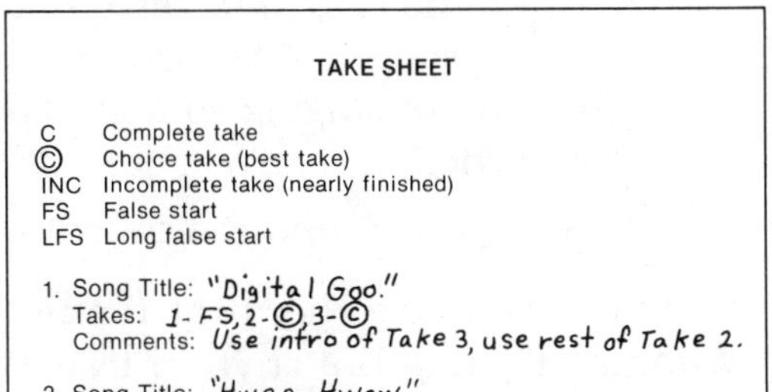

Figure 9-3.
Take sheet.

TAKE SHEET

C Complete take
Ⓒ Choice take (best take)
INC Incomplete take (nearly finished)
FS False start
LFS Long false start

1. Song Title: "Digital Goo."
   Takes: 1-FS, 2-Ⓒ, 3-Ⓒ
   Comments: Use intro of Take 3, use rest of Take 2.
2. Song Title: "Hwee-Hwow."
   Takes: 1-FS, 2-FS, 3-LFS, 4-INC, 5-FS, 6-INC, 7-LFS,
   Comments: Forget it!
3. Song Title: "Busted Pan-Pot Blues."
   Takes: 1-C, 2-FS, 3-Ⓒ, 4-C
   Comments: Mono.

headphones and sings along with them, a procedure followed for all subsequent overdubs. You record the overdub either on an open track or on a track you don't mind erasing. Here's the procedure:

1. Set up a mic for the vocalist and plug it into the mixer. Relabel the inputs if necessary.
2. Connect the vocalist's headphones to a small power amp driven from the "cue-mix" or "monitor-mix" output on the recorder-mixer.
3. Set the "monitor select" switch to "bus" for the vocalist's track; set it to "tape" for the other tracks.
4. Set the recorded tape tracks to "sync" or "safe" mode and set the track to be recorded to "record ready" mode. Assign the vocal to an unused tape track; turn up its fader; and adjust the input attenuation, level, and equalization as needed.
5. Set the tape counter to 000 at the beginning of the song.
6. Play the tape and use the "monitor-mixer" ("tape-cue") knobs to blend the previously recorded tracks with the live vocal-microphone signal. This creates a "cue mix"—the balance among instruments and vocals heard through the studio headphones. For example, suppose you've already recorded drums, bass, and guitars, and you're ready to add a vocal. While monitoring the cue bus, play the tape and set up a cue mix so that the vocalist can hear her- or himself and the previously recorded tracks through the headphones. Remember, the cue mix is not going on tape—it just allows you to hear how the vocalist blends with the other tracks.
7. Using the return-to-zero function, rewind to the beginning of the song and record the vocal. The singer must be quiet when not singing so that extraneous noises are not recorded.
8. Hit "stop" when the overdub is done.

If a mistake is made, the vocal track (or any overdub) can be rerecorded without affecting the rhythm tracks. Other overdubs might include harmony vocals or instrumental solos. You can even redo rhythm tracks that were unsatisfactory in sound or performance at the original recording session.

If an overdub occurs only in the middle of a song you don't need to rewind to the beginning. Set the tape counter to 000 a few seconds before the overdub point, and use the return-to-zero feature to practice the overdub.

Punching In

You might want to correct musical errors in a track by punching in a new passage. Here's how:

1. Set the tape counter to 000 a few seconds before the part needing correction.
2. Play the tape track to the musician over headphones.
3. During a rest (a pause in the track) just before the part needing correction, punch in the record button (or use a footswitch plugged into the recorder-mixer). If necessary, the musician can signal you where to punch in by pointing at you or by using the footswitch. Record the corrected musical part.
4. Immediately after the corrected part is performed, punch out of "record" mode (or use the footswitch) to avoid erasing the rest of the track.
5. Using the return-to-zero function, rewind the tape and play it back. If necessary, you can rerecord the punch.

Often it's difficult for a musician to get all the way through a long difficult solo or musical line without making a mistake. In this case, you can punch in and out to record the part in successive segments.

With care, you can punch in additional instruments on a completely full tape, by recording them in the pauses on previously recorded tracks. For example, suppose all the tracks are full but you want to add a cymbal roll at the beginning of the chorus. Find a track that has a pause at that moment, and punch in the cymbal roll there.

Bouncing Tracks

What if you want to bounce tracks? Let's say you want to bounce or copy tracks 1, 2, and 3 to track 4. If your machine has "record/play/send" switches, proceed as follows:

1. Monitor only track 4.
2. Set the "record/play/send" switches to "send" for input modules 1, 2, and 3. Set the switch to "play" for input module 4.

3. Play the tape.
4. Use the output knobs in input modules 1, 2, and 3 to mix the tracks. Set the recording level on track 4 to peak around 0.
5. When you're happy with the mix, rewind to the beginning of the song.
6. Set the "record/play/send" switch for track 4 to "record."
7. Start recording. Tracks 1, 2, and 3 will be mixed and recorded onto track 4.
8. Hit "stop" just after the end of the song.

Tracks 1, 2, and 3 still contain their original program. You can erase them either by recording no signal on these tracks or by recording new overdubs over them.

If your machine *does not have* "record/play/send" switches, proceed as follows:

1. Monitor only track 4.
2. Assign input modules 1, 2, and 3 to track 4.
3. Set all tracks to "play" or "safe" mode.
4. On input modules 1, 2, and 3, set the input-selector switches to "tape."
5. Play the tape.
6. Using input faders 1, 2, and 3, mix the tracks and set the recording level on track 4 to peak around 0.
7. When you're happy with the mix, rewind to the beginning of the song.
8. Set track 4 to "record ready" mode.
9. Start recording. Tracks 1, 2, and 3 will be mixed and recorded onto track 4.
10. Hit "stop" just after at the end of the song.

Tracks 1, 2, and 3 still have their original program on them. Either record new overdubs over them or erase them by recording no signal on these tracks.

While bouncing, you can add a live mic signal by plugging it into input 4 and assigning input 4 to track 4. In this way, you can record up to 10 tracks on a 4-track machine while bouncing tracks (Table 9-2). Here's how:

Table 9-2. Bouncing procedure to record 10 tracks with a 4-track recorder.

Step A	Step B	Step C	Step D	Step E	Step F	Step G
Track	*Track*	*Track*	*Track*	*Track*	*Track*	*Track*
1 A*	1 —	1 E*	1 —	1 H*	1 —	1 J*
2 B*	2 —	2 F*	2 —	2 —	2 HI*	2 HI
3 C*	3 —	3 —	3 EFG*	3 EFG	3 EFG	3 EFG
4 —	4 ABCD*	4 ABCD	4 ABCD	4 ABCD	4 ABCD	4 ABCD

Each track with an asterisk is a live microphone signal. Tracks without an asterisk are already recorded on tape.

1. Record three instruments on tracks 1, 2, and 3 (Step A).
2. Mix these three tracks with a live instrument signal and record the result on track 4 (Step B).
3. Record two more instruments on tracks 1 and 2 (Step C).
4. Bounce tracks 1 and 2 onto track 3 while mixing in another live instrument (Step D).
5. Record one more instrument on track 1 (Step E).
6. Bounce track 1 to track 2 while mixing in another live instrument (Step F).
7. Record one more instrument on track 1 (Step G).

You can continue this process to add even more instruments, but every rerecording adds noise, distortion, and frequency-response errors. This is called *generation loss.*

Try a similar procedure for other track-bouncing combinations, setting up your recorder-mixer appropriately for the desired track assignments.

Mixdown

After all four tracks are recorded, it's time to mix or combine them to 2-track stereo. You prepare the mixer and tape decks, erase unwanted sounds, and play the multitrack tape through the mixer while adjusting balances, panning, equalization, reverberation, and effects. Once you've rehearsed the mix to perfection, you record it onto a 2-track recorder. Let's look at these procedures in detail.

Set Up the Mixer

To begin, on the back of the unit locate the jacks for output channels 1 and 2 (they might be called "Bus 1 and 2" or "Stereo mix bus"). Plug these outputs into the line or aux inputs of a stereo cassette recorder or open-reel deck.

Clean and demagnetize the tape machines. Set all the mixer controls to "off," "flat," or "zero."

Tape a strip of paper along the front of the mixer to write which instrument(s) each fader affects. Keep this strip with the multitrack master tape so that you can use it each time the master is played.

Set the input-selector switches on the mixer to "tape" (or "track" or "remix") because you'll be mixing down the 4-track tape. Monitor the 2-track stereo mix bus.

Set the master fader(s) about ¾ up (at the shaded portion of fader travel).

Assign each track to channels 1 and 2, and use the pan pots to place each track where you want them between your stereo speakers. Typically, bass, kick drum, and vocals go to center and keyboards and guitars can be panned left and right or half-left and half-right.

You may want to pan extreme left and right tracks slightly toward the center; this keeps the tracks from sounding too isolated in space. Try not to pan everything to the middle—you'll wind up with a mono tape.

Erase Unwanted Program Material

Play the multitrack tape and listen to each track alone. Erase unwanted sounds and outtakes so that you won't be surprised during the mixdown. Examples of unwanted sounds are coughs, comments, and guitar string noises heard before the guitarist starts playing. You may want to erase entire tracks or segments that don't add to the song. If a noise occurs just before the musician starts playing, erasing the noise may accidentally erase the musical part. To prevent this, turn the tape upside down by reversing the reels, then find the track of the desired instrument playing backwards. Play the tape section that comes just after the noise. You'll hear it playing in reverse. Just after the reverse part ends, punch that track into record mode to erase the noise.

Compress the Vocal Track

The vocal track might be too loud or too quiet relative to the instruments, because vocals have a wider dynamic range than instruments. If this occurs in a few spots only, you can compensate by adjusting the vocal fader during the mixdown.

In worse cases, you may want to run the vocal track through a compressor patched into the access jacks for the vocal input module, if these jacks are available. Otherwise patch it between the vocal tape-track output and a mixer tape-track input. The compressor will keep the loudness of the vocals more constant, making them easier to hear throughout the mix.

If you compress a track during mixdown, it will become a little noisier, but at least you can change compressor settings as needed to suit the mix. If you compress a track while recording it, that track will not become noisier during mixdown, but you'll have to decide on compressor settings in advance.

Adjust the Faders

Use the input faders to adjust the volume of each track for a pleasing balance, similar to what you hear on records. You should be able to hear each instrument clearly. In some inexpensive units, you use the monmix controls for this function. As a starting point, you may want to set the mix so that all the instruments and vocals sound equally loud, then turn up the most important tracks and turn down background instruments. Or you can bring up one track at a time and blend it with the other tracks. For example, first bring up the kick drum, then add bass, and balance the two together. Next add drums, guitars, keyboards, then vocals.

To reduce tape noise, mute any track that has nothing playing at the moment. That is, if there is a long silence during a track, mute that silent portion. Mute unrecorded tracks as well.

Adjust Equalization

Next, adjust the equalization for the desired tonal balance on each track. Normally you leave EQ flat (knob set at 12 o'clock). But if a track sounds too dull, turn up the treble or high frequencies. If a track sounds too bassy, turn down the bass or low frequencies, and so on. Many directional microphones (such as cardioids) boost the bass when placed a few inches from their sound source, resulting in an unnaturally bassy sound. You can control this by turning down the bass (low-frequency) EQ until the sound is natural. Presence or definition is often enhanced by boosting frequencies around 5 kHz (1.5–2.5 kHz for bass instruments).

In most recorder-mixers, each track can be equalized independently. If cymbals and vocals are recorded on separate tracks, for example, you can add crispness (high frequencies) to the cymbals without affecting the tone quality of the vocals. Or you can remove a "tubby" tone from the kick drum without affecting the keyboards.

Add Effects

You might want to plug in an external reverb or delay unit to add spaciousness to the sound. This device connects between the aux-send and -receive jacks. If your recorder-mixer has an aux-return control, set it halfway up. Using the aux knobs on the mixer, adjust the amount of delay or reverb for each track as desired. Don't overdo the reverb, or the sound will become muddy. Normally, vocals and lead guitar get the most reverb, while bass and kick drum get little or no reverb so that they can retain their clarity.

Set Recording Levels

As you're mixing, be sure to keep the recording level peaking around 0 VU (+3 VU maximum) by adjusting the input faders, with the master fader(s) remaining about ¾ up. The meters for channels 1 and 2 in the mixer and in the 2-track recorder should all peak around 0 VU.

Fine-Tune the Mix

If level changes are required during the mixdown, mark on the faders the settings for each change.

Make a *cue sheet* that notes the mixer changes required at various tape-counter times. For example:

- 1:10 Bring up lead-guitar solo to +3 dB
- 1:49 Drop lead guitar to −2 dB.
- 2:42 Add +6 dB at 12 kHz to synthesizer bell effect.
- 3:05 Start fade, out at 3:15.

Don't mix in too much bass. Sometimes it's hard to tell how much bass is appropriate, owing to variables in the monitoring system. Typically, the bass is mixed in at about −8 to −4 VU when metered alone, but you normally mix by ear, of course, rather than by watching the meters.

Using the tape recorder's return-to-zero function, play the multitrack tape several times to perfect and practice the mix. Set your balances, equalization, and effects. It's a good idea to play a record with tunes like those you're recording to hear a typical mix for that style of music.

When you mix, listen briefly to each instrument, in turn, and to the mix as a whole. If you hear something you don't like, fix it. Is the vocal too tubby? Roll off the bass on the vocal track. Is the kick drum too quiet? Turn it up. Is the lead-guitar solo too dead? Turn up its aux send. Tips on judging the sound quality of the recording are given in Chapter 11.

Record the Mix

Once everything is set the way you want it, put the 2-track recorder in record mode and record the mix.

If you want to fade out the end of the tune, slowly pull down the master fader. Start the fade relatively quickly, then slow down as you fade. The slower the song, the slower the fade should be.

The mixdown is now complete. Repeat this procedure for all the best takes on the multitrack reels.

Summary of the Mixer Operating Procedures

The following lists are step-by-step procedures you will use. They are summarized here for easy reference.

Recording

1. Turn up the headphone or monitor volume control.
2. Assign the instruments to their tracks.

3. Turn up the submasters, bus masters, or group faders (if any) and the master fader to design center (the shaded portion of fader travel, about ¾ up).
4. Adjust the input attenuators ("trim").
5. Set the submixes and recording levels.
6. Set the cue-monitor mix.
7. Record onto the multitrack tape.

Overdubbing

1. Assign the instruments or vocals to be recorded to open tracks. An open track is blank or has already been bounced.
2. Turn up the cue-monitor system.
3. Turn up the submasters and master to design center.
4. Play the multitrack tape in sync mode and set up a cue-monitor mix.
5. While a live musician is playing, adjust his or her input attenuation and recording level.
6. Set the cue-monitor mix to include the sound of the instrument or vocal being added.
7. Record the new parts on open tracks.
8. Punch in and bounce as needed.

Mixdown

1. Set the input selectors to accept the multitrack tape signals.
2. Monitor channels 1 and 2 (2-track mix bus).
3. Assign tape tracks to channels 1 and 2.
4. Turn up the master fader to design center. In some mixers, the submasters also should be up.
5. Set a rough mix with the input faders.
6. Set equalization, reverberation, and effects.
7. Perfect the mix and set the recording levels.
8. Record onto the 2-track tape.

Assembling the Master Reel

If you're using an open-reel 2-track recorder for mixdown, you're now ready to assemble that tape into a finished format for duplication. It will contain the songs in the desired order, plus leader tape between songs.

Leader Length

The length of leader between songs depends on how long a pause you want between them. Four seconds is typical. Use longer leader if you want the listener to get out of the mood of the piece just heard before going on to the next. Use shorter leader either to change the mood abruptly or to make similar songs flow together.

Splice several seconds of leader at the beginning and end of your program.

Use a piece of masking tape to fasten the leader tail to the reel and print "TAIL OUT" on the masking tape. Type or print a neat label for the tape reel including title, artist, date, and the words "stereo mix master."

Labeling

Include the following information on the tape-box label:

1. For an open-reel master tape, note the tape-head format (usually half-track stereo), stereo/mono, tape speed, and "tail-out" designation.
2. Type of noise reduction used.
3. Demo title, artist, song titles.

Conclusion

Congratulations! There's your finished master tape, ready to play for others. It's amazing how the long hours of work with lots of complex equipment have concentrated into that little tape. But it's been fun. You have crafted a product you can be proud of. When played, it will recreate a musical experience in the ears and the mind of the listener—no small achievement.

It's a good idea to make a safety copy of the master tape in case it is lost or damaged. Take care in setting recording levels while copying: at the loudest part of the tape, the VU meter of the copying deck should peak at +3 VU maximum (or 0 dB on an LED bargraph meter).

For small quantities of cassettes, you may want to copy the master tape yourself as many times as needed. You'll soon have a number of quality demo tapes that you can send to prospective clients or record companies.

10 On-Location Recording of Popular Music

Rather than record your demo tape in your home, you may want to record a live performance in a club or concert hall. Many bands want to be recorded in concert because they feel that's when they play best. Your job is to capture that performance on tape and bring it back alive.

There are many ways to do this. We'll start with simple two-microphone techniques and work our way up to elaborate multitrack setups. Then we'll step through an on-location session.

Monitoring

Headphones, rather than loudspeakers, are generally used for on-location monitoring because they are more portable and provide consistent sound in different environments. Plus, headphones—particularly the closed-cup, over-the-ear type—partly block out the live music so you can better hear what's going on tape.

If you record a band in the same room, the live sound of the band will leak through the headphones' ear seal, making it hard to hear the monitored signal clearly, especially the bass. This might not be a problem if you're recording a relatively quiet acoustic group. But to clearly monitor a loud rock group within the same room, you'll need to adjust the headphones to quite loud; this might damage your hearing.

The preferred practice for recording louder groups is to set up your equipment in a separate room. Then run some microphone extension cables (or a snake) from your mixer out to the performance area. Close the door, slip on the headphones, and monitor the sound. You'll be able to hear more clearly without the danger of hurting your ears. During intermissions, you can play back the tape to hear what you've just recorded.

Recording with Two Microphones

A beginning recordist might start with two microphones and a 2-track tape deck, which is the easiest method of recording a group. Small acoustic ensembles often can be well recorded this way. With rock groups, however, we're accustomed to a clean, tight recorded sound picked up by multiple closely placed microphones—a sound you can't duplicate with a simple two-microphone pickup. However, a two-mic recording is useful for musicians who want to hear how they blend in the audience area. It might even be adequate as an audition tape for local jobs. Recording this way is much simpler, faster, and cheaper than multi-mic, multitrack recording. Still, if time and budget permit, you'll get better sound with a more elaborate setup.

Two Crossed Cardioid Microphones

Here's one method of recording with two microphones. First, mount two high-quality cardioid microphones on a stereo microphone-stand adapter (or improvise a holder with duct tape). Angle them 110° apart (55° to the right and left of center) and space their grilles 7″ apart horizontally. This is the *Office de Radiodiffusion-Television Française* (ORTF) stereo miking system.

You also could use a stereo microphone to achieve the same effect. Place the stereo mic about 3′ in front of a folk group or vocal quartet or 10–15′ in front of a rock group on stage. Use a microphone stand or hang the mics out of the reach of the audience.

Two Spaced Microphones

Most rock groups use loudspeakers at each end of the stage to reinforce the vocals (and often certain instruments). A centrally placed stereo pair of microphones, being far from the sound-reinforcement speakers, may not pick up the vocals adequately. For better control over the vocal-instrumental balance, try aiming two cardioid microphones straight ahead toward the group, spaced about 5–15′ apart, as shown in Figure 10-1. Placing the microphones far apart (that is, close to the sound-reinforcement speakers) makes the vocals louder in the recording. Placing the mics closer makes the vocals quieter. The stereo imaging of this arrangement is poorer than with the ORTF system, but at least you can control the balance between instruments and vocals.

If you're recording a small acoustic group, try the methods suggested in Chapter 4.

Preventing Mic-Preamp Overload

If the playback sounds distorted even though you did not exceed a normal recording level, the microphones probably overloaded the microphone preamplifiers in the tape deck. A microphone preamp is a circuit in the tape recorder that amplifies the weak microphone signals to a usable level. With loud sound sources such as rock groups, a microphone can put out a signal strong enough to cause distortion in the mic preamp.

Some decks include a pad or input attenuator, which reduces the microphone signal level before it reaches the preamp, thereby preventing distortion. Others have a high-impedance microphone input, which will act as an attenuator if used with a low-impedance microphone. Some condenser microphones have switchable internal pads that reduce distortion within the microphone. You or a friend who is good with electronics can build a pad with 20-dB attenuation as shown in Figure 10-2, or you could buy some plug-in pads from your microphone dealer. If you have to set your record-level controls very

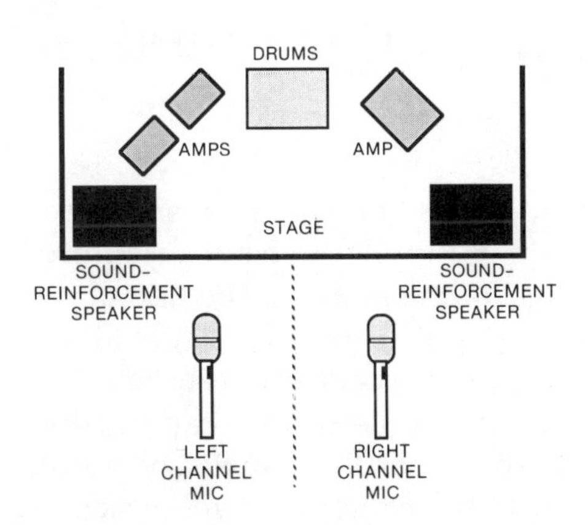

Figure 10-1. Recording a rock group with two spaced microphones.

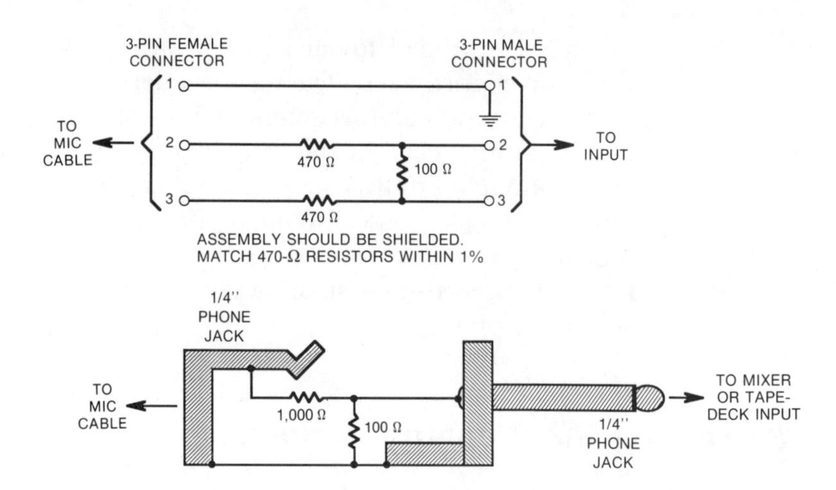

Figure 10-2. Balanced and unbalanced microphone pads.

low (less than ⅓ up) to obtain a 0-VU recording level, that's a good indication you need to use a pad.

Recording

The actual recording process is simple. Hit the record button(s) and set the recording levels to peak around 0 maximum on your cassette-deck meters (+3 VU for an open-reel deck). Once the levels are set, leave them alone as much as possible. If you must change them, do so slowly and try to follow the dynamics of the music.

Bring more than enough tape for all the songs you want to record. Switch tapes during pauses or intermissions.

Recording from the Sound-Reinforcement Mixer

Sometimes you can get a good recording simply by plugging into the main output of the band's reinforcement mixer. Connect the line output(s) of the mixer to the line or aux input(s) of your 2-track recorder. Use the mixer output that is ahead of any graphic equalizer used to correct the speakers' frequency response, as shown in Figure 10-3.

Note: Some mixers can produce a signal whose level is too high for the recorder's auxiliary input, which causes distortion. This is probably occurring if your record-level controls have to be set very

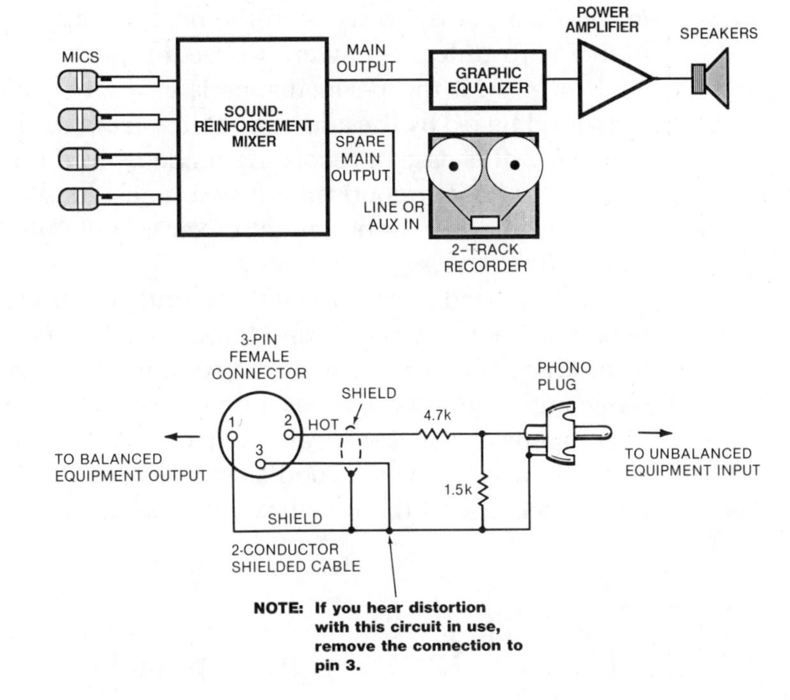

Figure 10-3. Recording from the sound-reinforcement mixer.

Figure 10-4. 12-dB pad for matching a balanced +4-dBm output to an unbalanced −10-dBV input.

low. To reduce the output level of the mixer, turn it down so that its signal peaks around −12 VU on the mixer meters, and turn up the PA power amplifier to compensate. Alternatively, make a 12-dB pad as shown in Figure 10-4. The output level of a balanced-output mixer is called +4 dBm or +4; the input level of unbalanced equipment is called −10 dBV or −10.

Why is there a 12-dB difference between +4 dBm and −10 dBV? Well, +4 dBm is 1.23 V, and −10 dBV is 0.3 V. The second voltage is 12 dB below the first.

Drawbacks

Recording from the band's mixer works best when all the instruments are miked and mixed through that mixer. The recorded mix still might be bad, however, especially if the room is small-to-medium in size. Here's why: the operator of the band's mixer hears a combination of the live sound of the band and the reinforced sound through the house system, and tries to get a good mix of both elements. That means the signal is mixed to augment the live sound, not to sound good by itself. A recording made from the band's mixer is likely to sound too strong in the vocals and too weak in the bass. It's a compromise you'll have to live with unless you want to use a separate recording mixer.

However, if the performance is in a large hall or arena, most of the sound heard by the audience comes from the sound-reinforcement system. In this case, a recording made from the reinforcement mixer is likely to have a good mix: it will be as good as the live mix is. To record from the reinforcement mixer, simply plug in, hit record, and watch your levels.

This method works best if the sound-reinforcement speakers were previously equalized to sound "hi-fi" when playing a good recording. If the frequency response of the reinforcement speakers is not wide-range and smooth, the mixer operator may equalize each instrument to compensate for the speakers. If you record this compensated mix and play it back over a good stereo system, the tonal balance will be wrong because of the equalization used on the reinforcement mixer.

Recording Vocals from the Sound-Reinforcement Mixer

In some small systems, only the vocals are reinforced. You can take a line-level signal from the band's mixer for the vocals and use your own microphones and direct boxes for the instruments. To do this, you'll need a separate mixer for recording. Figure 10-5 shows the connections.

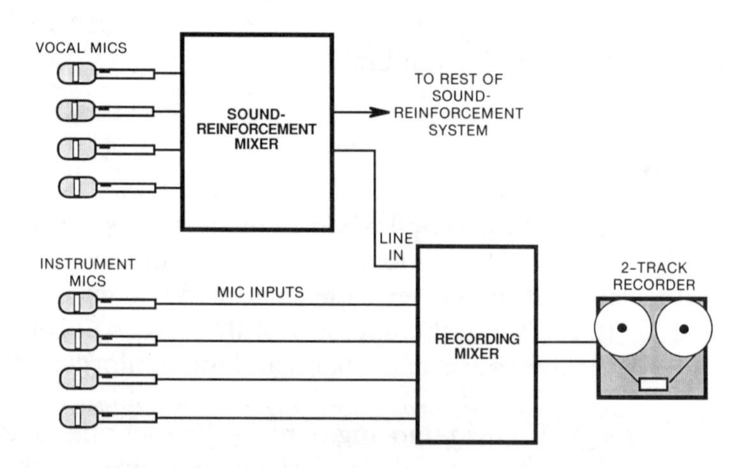

Figure 10-5. Recording vocals from the reinforcement mixer, with separate mics for instruments.

For on-location work, you must place each microphone within a few inches of its source to reject feedback, leakage, and room acoustics. See Chapter 7 for some recommended microphone techniques.

Check the input-overload LEDs (clip lights or peak lights) on the band's mixer (if any) and on your recording mixer to make sure the vocal mics aren't overloading their inputs. Now you're ready to mix. Listen with headphones and mix the microphone signals with the vocal signal from the band's mixer. This procedure is much like that of a multitrack mixdown, except that you're mixing live as the band plays. Using faders, equalization, panning, and effects, try to make the mix sound like records or CDs you've heard through those headphones. It helps to be familiar with the music being performed. If you're making an audition tape, use effects and reverb sparingly or not at all, because this tape should represent how the band sounds live.

Ambience Microphones

If you have enough microphone inputs, you can add one or two ambience microphones to pick up the room acoustics and audience sounds. This helps the recording sound "live" because without ambience microphones, the recording might sound too dry, as if it were done in a studio.

One popular technique is to mount two boundary microphones or miniature omni condenser mics on the walls or ceiling; they can provide a clear, realistic pickup of audience reaction. Alternatively, hang two crossed cardioids or spaced omnis over the audience.

Ambience microphones can muddy the sound if mixed in too loudly. Keep them down in level, just enough to add some "atmosphere." Bring them up gently to emphasize crowd reactions.

Splitting the Microphones

As we've seen, a good house mix does not guarantee a good recording mix. It's better to make an independent recording mix by using a separate mixer and separate microphones.

Y-Adapter

The stage will be cluttered if you place a recording microphone next to every reinforcement microphone. It's especially clumsy to double the vocal mics. Instead, you can plug a Y-adapter (shown in Figure 10-6) into the end of each vocalist's microphone cable. This adapter splits the microphone signal two ways: to the reinforcement mixer and to the recording mixer.

Plug one output connector of the Y into a cable going to a reinforcement-mixer mic input. Plug the other output connector of the Y into a cable going to a recording-mixer mic input.

Powering Precautions for Y-Adapter Use

To prevent hum, be sure to plug both the recording mixer and the sound-reinforcement mixer into the same outlet strip so you get ac power from the same place that the sound-reinforcement mixer does. Run a long, thick (14- or 16-gauge) extension cord from that point to the recording mixer. Plug a multiple ac outlet strip into the extension cord, then plug all your equipment into the outlet strip.

Be sure that your ac power source is not shared with lighting dimmers or heavy machinery; these devices can cause noises or buzzes

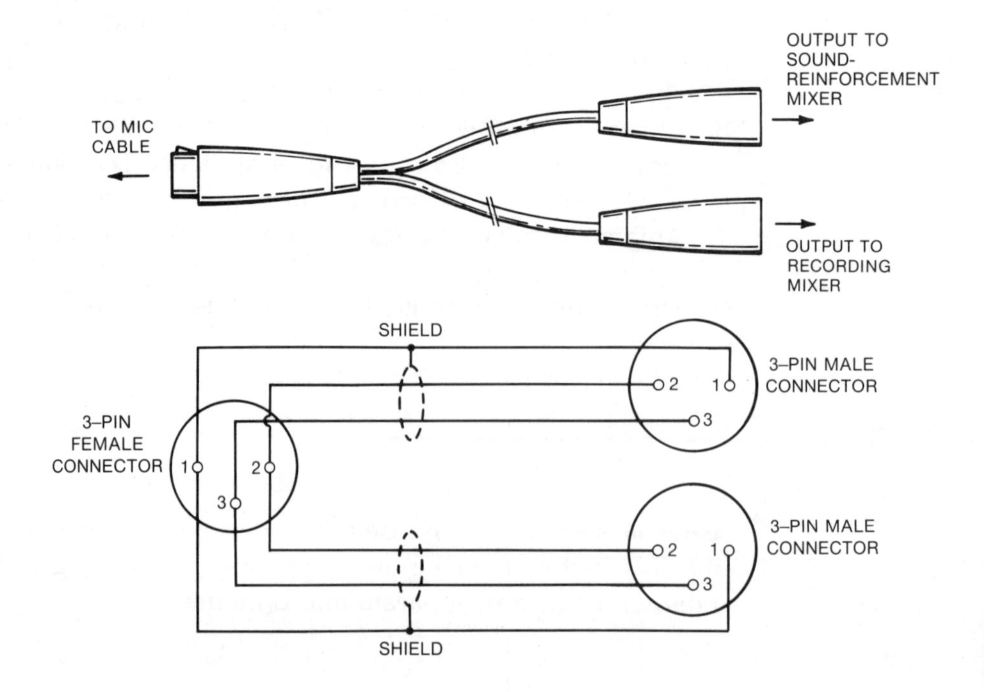

Figure 10-6. A Y-adapter for splitting microphone signals.

in the audio. If you use phantom powering, supply it from one console only.

Microphone Splitter

You're less likely to encounter hum if you use a microphone splitter (shown in Figure 10-7), which uses transformers to isolate the mixers. They are available at sound dealers and some music stores, but, unfortunately, are expensive and you need one split channel for every microphone you want to share with the sound-reinforcement system.

In some sound-reinforcement systems, every instrument is miked with a high-quality microphone. Then you can split all the microphones. But in many other systems, either the vocals only are miked for reinforcement or you prefer not to use the band's instrument mics. In that case, split the vocal mics and use your own recording mics on the instruments.

Recording Live to 2-Track

A recording mixed live to 2-track can sound as good as a commercial LP of a live concert. You bypass the noise and distortion added by a multitrack recorder in this way, but the mix may not be optimum because you have to mix as the musicians are playing. A multitrack recorder, on the other hand, lets you tailor the mix after the concert.

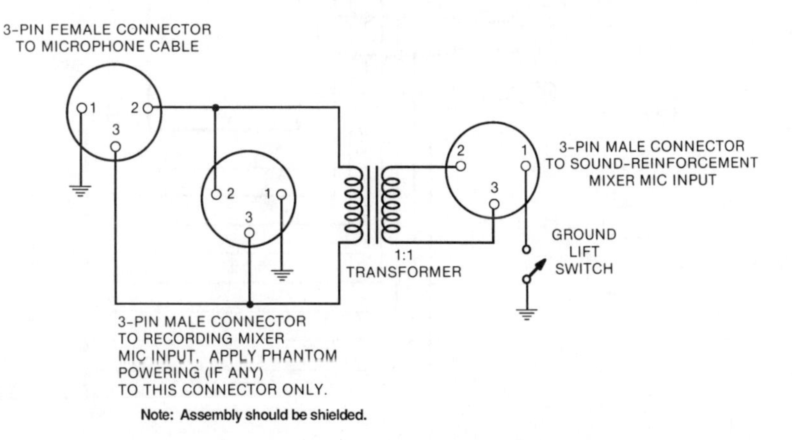

Figure 10-7. Transformer-isolated microphone splitter.

Multitrack Recording

Now we're getting into professional techniques. Each microphone on stage is split to feed the sound-reinforcement mixer and a separate multichannel recording mixer, as diagrammed in Figure 10-8. Some splitters have a third output in order to feed a stage-monitor mixer as well. To prevent hum caused by interconnecting the three systems, ground the microphone-cable shields to the recording mixer only. "Float" (disconnect) the cable shields going to the house mixer and monitor mixer with ground-lift switches on the splitter.

Each microphone, or each instrument's group of microphones, is assigned to a separate track of a multitrack recorder. After making the recording, you mix down the tracks in your home studio, spending as much time as needed to perfect the mix. You can even overdub parts that were flubbed during the live performance, taking care to match the overdubbed sound to the original recording.

Figure 10-9 shows a typical equipment layout for an on-location recording of a rock group. Three systems are in use—sound-reinforcement, recording, and monitor mixing—and the mic signals are split three ways to feed these systems.

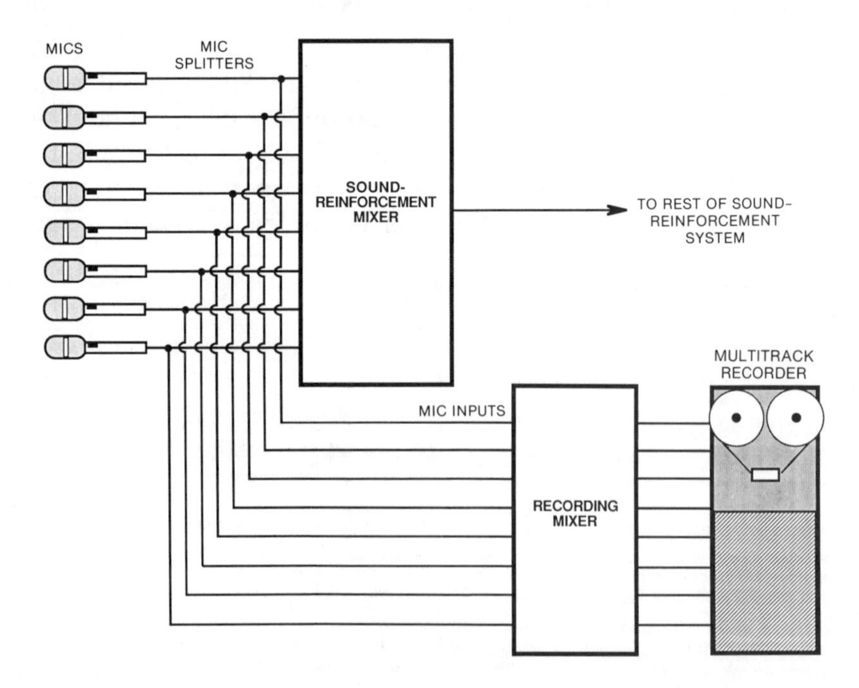

Figure 10-8. Recording with mic splitters into a multitrack recorder.

Figure 10-9. Typical layout for an on-location recording of a rock group.

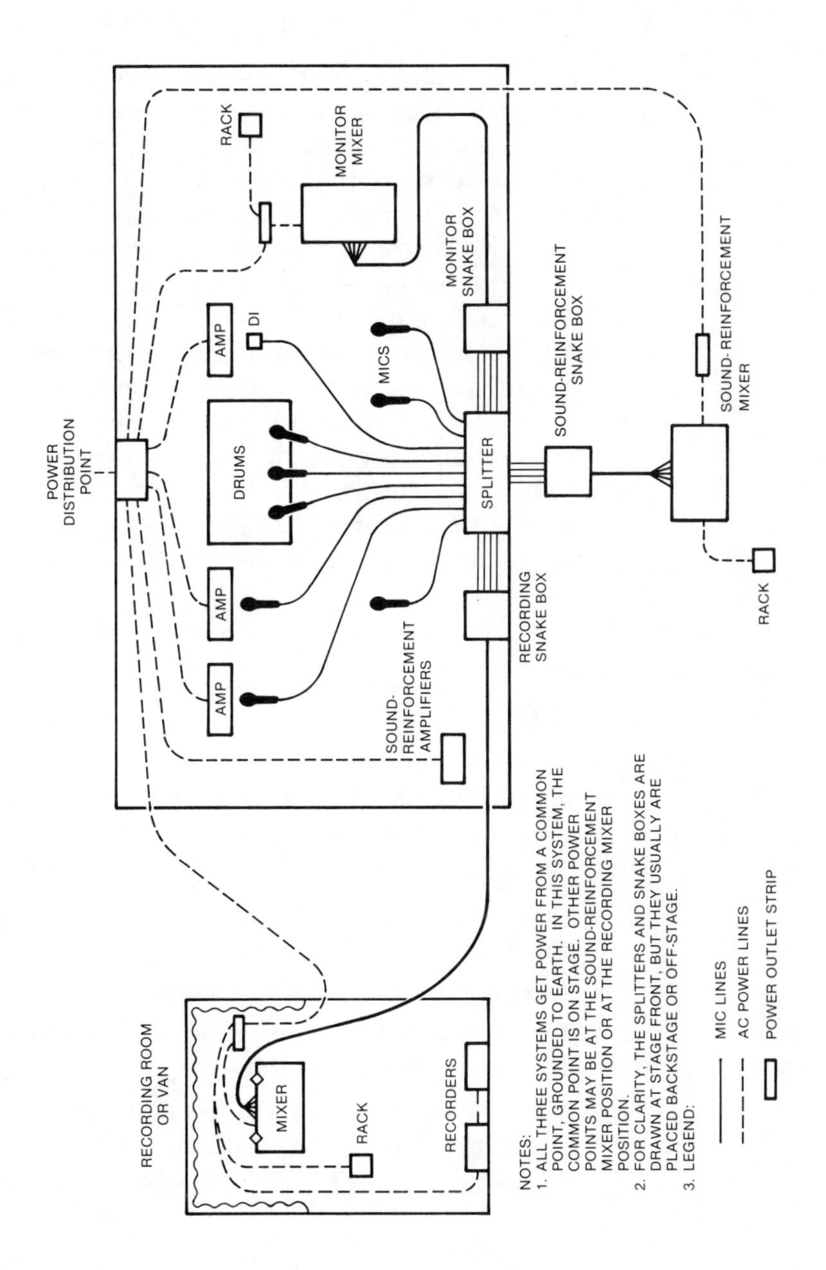

Track Formats

The 4-track format is probably the most difficult to use for live recording. That's because you have to sub-mix several microphones onto each track and monitor all four tracks. You must set up the 4-track monitor mix very carefully, because you can't change the mix within each track after recording (except slightly with equalization).

A jazz trio might provide the easiest 4-track recording situation. You could put the bass on track 1, a drums mix on track 2, and stereo piano on tracks 3 and 4.

Eight tracks are easier to work with because you do most of the mixing after the concert. You might need to mix the drum microphones to one or two tracks at the recording session, but typically each microphone feeds its own track. Most of your work during recording is setting levels.

Even the 2-track format is easier than 4-track. Monitoring is simple; the mix you hear during recording is the mix you'll hear during playback.

Summary of Techniques

We've covered a variety of on-location recording techniques. In general, the more sophisticated the setup, the better the sound. Here's a list of the methods discussed, from simple to complex:

- Place two microphones out front; use with pads into a 2-track tape deck.
- Record from the sound-reinforcement mixer.
- Record vocals from the sound-reinforcement mixer and mike the instruments separately.
- Use microphone splitters. Mix all the microphones with a recording mixer live to 2-track.
- Record on a multitrack tape machine for later mixdown.

Miscellaneous Tips

So far we have overviewed various on-location recording methods. This section offers suggestions on doing the job more effectively.

- Go to the club or hall before the concert so that you can plan where to put your equipment, how much cable is needed, where people are likely to trip over cables, and so on.
- Draw a block diagram of your recording system to generate an equipment list. Check each cable and piece of equipment for proper operation before going on location. Check off the equipment from the list as you pack.
- In very cold weather, don't pack your recorder or batteries in the car until just before you leave. Otherwise, the recorder lubrication may become stiff and the batteries may lose some charge.
- In hot weather, don't leave tapes in the car because the heat might partially erase the signal on tape or cause print-through.
- Arrive several hours ahead of time for setup. Expect failures—something always goes wrong—or the unexpected. Allow more time for troubleshooting than you think you'll need.
- Be sure to include spare tape reels, spare cables, hub adapters, pencil and paper, guitar cords, and guitar strings. Bring a tool kit with screwdrivers, pliers, soldering iron, solder, duct tape, connectors, audio adapters, electrical 3-to-2 adapters, spare cables, a flashlight, and 9-V batteries.
- To prevent hum and ultrasonic oscillations, avoid bundling microphone cables, line-level cables, and power cables together. If you must cross mic cables and power cables, do so at right angles and space them vertically.
- Don't leave a rat's nest of cables near the stage box. Coil the excess cable at each mic stand. That way, you can move the mics and reduce clutter at the stage box.
- Have an extra microphone and cable offstage in case a mic fails.
- Plug in one mic at a time and monitor it to check for hums and buzzes. Troubleshooting is easier if you listen to each mic as you connect it, rather than try to find a hum or buzz when all are plugged in.
- Check and debug one system at a time: first the sound-reinforcement system, then the stage-monitor system, then the recording system, and so on. Again, this makes troubleshooting easier because you have only one system to troubleshoot.

- Overloud stage monitors can ruin a recording, so work with the sound-reinforcement operator toward a compromise. (Ask the operator to start with the monitors quiet, because the musicians always want them turned up louder.)
- Set recording levels during the sound check before the concert. It's better to set the levels a little too low than too high because, during mixdown, you can reduce noise but not distortion.
- If a concert will last longer than the running time of a reel of tape, switch reels at intermissions. Or, if you know which songs you want on the demo tape, record those only.
- After the gig, note equipment failures and fix broken equipment as soon as possible.

By following these suggestions, you should improve your efficiency—and your recordings—at on-location sessions.

A Sample On-Location Session

Can a professional, on-location recording be made with home recording equipment? You bet! Here is the story of one such recording, with examples of what to do—and what not to do.

A small group wants to make a live demo tape of their next club date. The major pieces of available equipment are:

- Fostex Model 80 8-track recorder (a small portable unit with Dolby C)
- Soundcraft 2-channel mixing console
- Revox A77 2-track tape recorder

I drew a block diagram of the recording system (shown in Figure 10-10) and used it to generate an equipment list (Table 10-1). Note that the list included small-but-necessary details such as adapters, razor blades, empty tape reels, and so on.

Each direct-connection cable was an earlier, simplified version of the one shown in Figure 2-5, without any high-frequency filtering capacitors. The cables took the signals from the guitar amps, synthesizer, and drum machine and sent them at a reduced level to the mixer mic inputs. One end of each cable plugged into the external-speaker jack of a guitar amplifier (or into a synth output); the other

Figure 10-10. Recording setup.

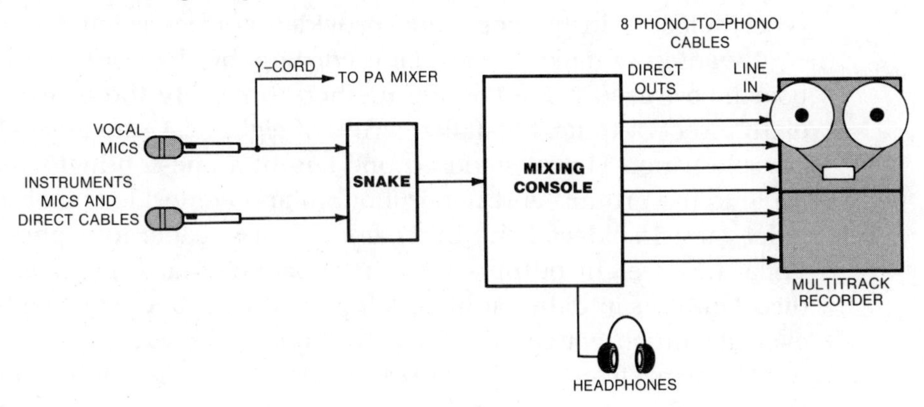

Table 10-1.
Equipment list.

Number	Equipment Part
2	Crown GLM-100 omni condenser mics
1	Direct box
1	Phone-to-phone cable for direct box
5	Direct-connection cables
15	Mic cables (including spares)
4	Y-cords (splitters)
1	Snake
1	Soundcraft mixing console
1	Sony MDR-V6 headphones
8	Phono-to-phone cables plus 2 spares
1	Fostex Model 80 8-track recorder
8	Reels of Ampex 456 blank tape
2	Empty takeup reels
1	Cleaning and repair kit (alcohol, swabs, tools, razor blades, adapters, etc.)
1	Notebook and pen
1	Roll duct tape
1	Roll masking tape
1	Felt-tip marker
1	Power outlet strip
1	Power extension cord
1	Flashlight

end plugged into a mixer mic input. It was an inexpensive way to record electric instruments and provided perfect isolation.

I wanted to make an 8-track recording, but had only a 2-channel board. So I used the following method to modify the board to have eight direct outputs. I installed a row of eight RCA phono jacks in the console panel. Then I soldered one end of a one-conductor shielded cable to the ground and wiper (bottom and center) terminals of each fader, and I soldered the other end of each cable to a phono jack. Voila: I had eight outputs from a 2-channel board. I plugged these direct outputs into the multitrack inputs. If your budget permits, however, it's much better to use a 4- or 8-channel board.

The signal from each fader was line level (−10 dBV), which matched the input requirements of the Fostex multitrack recorder. Each fader controlled the recording level of the track to which it was connected.

Presession Planning

The first question to ask a band you want to record is: "What is the instrumentation?" (What are the instruments you'll be playing, and how many vocalists will there be?) In this case, the list turned out as follows:

4	Vocals
1	Synth (mono output)
1	Drum machine (mono output)
1	Electric bass
1	Electric rhythm guitar
1	Set of percussion: congas and cowbell
1	Harmonica
1	Electric lead guitar (played through two amplifiers one at a time)
1	Acoustic guitar

How could all these instruments be fitted into eight tracks? Fortunately, not all the instruments were played at the same time. The bass player occasionally played synthesizer instead. The rhythm-guitar player sometimes switched to acoustic guitar. Also, the lead guitarist only played percussion or harmonica on some songs. As a result, the track assignments ended up like this:

Track 1. Bass or synth
2. Electric rhythm guitar or acoustic guitar
3. Drum machine
4. Electric lead guitar (guitar amplifier 1 or 2)
5. Percussion or harmonica
6. Vocal
7. Vocal
8. Vocal (2 singing into one mic)

Normally you'd use channel-assign buttons to send two instruments to the same track. But the 2-channel mixing board I was using did not have this feature, so I manually repatched the input signals according to what was being played at the moment.

Before leaving for the gig, I checked the operation of each piece of equipment—a must. Then, on the equipment list, I checked off each piece of equipment as I packed it.

On-Location Setup

At the club I found a spot off-stage to set up. Using a heavy extension cord, I connected my power outlet strip to the same outlet as the band's mixer. This helped to prevent hum caused by ground loops between the PA mixer and recording mixer. Then I plugged all my equipment into my outlet strip.

Next, I made connections between the mixer direct outputs and the multitrack recorder line inputs. I plugged the snake into the mixer and started connecting mics and direct cables.

The microphone list was as follows:

Electric bass: custom-made direct box plugged into the bass.
Synth, drum machine, electric guitars: direct cable.
Acoustic guitar, harmonica, percussion: 2 Crown GLM-100s.
Vocals: 3 Electro-Voice N/D 457s

I plugged each vocal mic into a Y-cord. One leg of the Y-cord fed the band's PA mixer; the other leg fed my recording snake. In other words, each vocal mic was split to feed two mixers. Normally a mic splitter with transformer-isolated outputs is best, but the Y-cords worked fine in this small setup.

The Crown GLM-100s are miniature omnidirectional condenser units. I taped one to the harmonica's PA mic to pick up either harmonica or percussion and taped another to the acoustic guitar's PA mic.

Finally, I plugged headphones into the mixer, cleaned the tape heads, and threaded on some blank tape.

Signal Check

It was time to see whether everything worked. Using the monitor knobs in the mixing board, I listened to each input by itself. One of the vocal mics sounded weak in the bass and low in level. I suspected that it was a high-impedance mic and that my mixing board's low-impedance input was loading it down. I substituted a low-impedance mic which worked fine.

After running guitar cables from the synth and drum machine to my mixer, I heard hum. So I used the direct-connection cable for each source. But then the signal levels coming from the drum machine and synth were low because the direct cable was designed for speaker-level signals, not line-level signals. Because of the weak signals, I found I had to push their faders all the way up. This is bad practice because it results in audible noise. So I asked the musicians to turn up the synth and drum machine to maximum. This is also bad practice! Musicians should not have to change their settings to accommodate your recording (unless the settings are way off).

After making this recording, I constructed some new cables to handle line-level signals.

Recording-Level Setting

During the sound check, I asked each musician to play alone so that I could set recording levels. To do this, I set each fader to design center (about ¾ up) and turned down the trim pot for each input until the LED overload indicator just stopped flashing. Some inputs needed no trimming, resulting in 0-dB recording levels on the multitrack machine.

When the gig was about to begin, I started recording and kept careful watch on the recording levels. Most of the levels needed retouching, but they were in the ballpark thanks to the sound check.

When a new instrument was going to be played, I switched input connectors and quickly reset recording levels. If you have channel-assign switches, it is better to plug each instrument's cable into a separate input, then assign the appropriate instrument to its track, and set the recording level for each. While the recording was in progress, I set up a rough monitor mix over headphones to approximate the finished mix, and to listen for any problems. Also, for each song, I kept notes of which instrument was on which track.

Since the Fostex Model 80 accepts only 7½″ reels and runs only at 15 ips, the recording time was about 22½ minutes per reel. The band knew about this limitation and paused while I switched reels. A few times, however, the musicians forgot to pause, so I missed the beginnings of those songs and had to fade them up during mixdown. This is not good practice. A bigger budget would have allowed a backup recorder connected in parallel with the first. I would have started the second machine just before the tape ran out on the first machine and edited the two tapes together back in the studio.

Mixdown

The next day at home, I set up the multitrack machine to play through the console line inputs (as is shown in Figure 10-11). The console main stereo outputs fed a Revox A77 2-track tape recorder. (For minimum noise and distortion, you'd use something like a DAT, a digital audio adapter into a VCR or a VCR with VHS Hi-Fi.) In-line with the cable between the console and the Revox was a 12-dB pad (shown in Figure 10-4). This reduced the console's high output level (called +4 dBm) to a lower level (called −10 dBV) suitable for the Revox input.

To add ambience to some of the tracks, I connected an Alesis Microverb to the console's aux-send and -return jacks.

I monitored the mixdown with a pair of Sony MDR-V6 headphones plugged into the Revox A77. Many engineers prefer to use loudspeakers for mixdown, but I'm used to mixing on headphones and, with

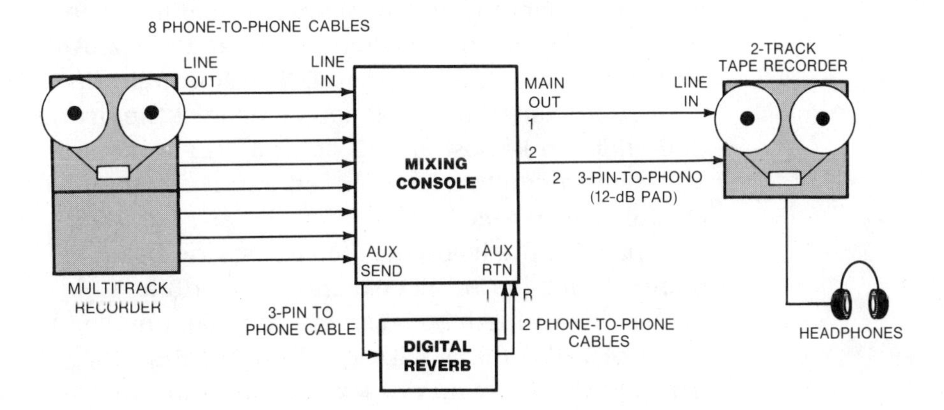

Figure 10-11. Mixdown setup.

them, don't have to worry about room acoustics or speaker deficiencies.

I threaded the first reel of tape on the multitrack. Referring to my track sheet, I made a designation strip—a long strip of paper placed below the faders, indicating which instrument each fader controlled. I muted unrecorded tracks.

I set all console controls to "off," "flat," or "zero" so they would have no effect. You have to start from ground zero in building a mix.

The first step in a mixdown is to listen to each track alone and clean it up: erase and filter out unwanted noises. The lead-guitar track had guitar-cord crackles which occurred before the guitarist started to play; I erased them.

The electric-guitar tracks were hissy in two ways: the rhythm-guitar amplifier was noisy, and the direct-connection cable picked up high-frequency hiss from the external speaker jack. In a guitar amplifier-speaker, the loudspeaker has a rolled-off high-frequency response that filters out amplifier hiss and dulls the sound of amplifier distortion, making it more listenable. But I was not recording the sound of the loudspeaker; I was recording the sound of the guitar amplifier. The direct-connection cable did not have a high-frequency filter, so it picked up the amplifier hiss.

Also, the guitar sound picked up by the direct-connection cable was too bright, or edgy, for the same reason. To compensate, I rolled off 12 kHz by 9 dB to restore a more natural tonal balance. The guitars also needed a low-frequency cut of 3 dB at 250 Hz and 80 Hz. Since making this recording, I have redesigned the direct-connection cable to include a filter that reduces high frequencies, as shown in Figure 2-5.

Next in the mixdown, I set the tape counter to 000 about 3 seconds before the song started. Time to roll tape. I hit "play" and brought up the drum-machine track, panned to center. The synthesized hi hat was a little dull, so I boosted +4 dB at 12 kHz. A 6-dB boost at 80 Hz added punch to the kick drum. Next I brought up the bass guitar (also panned to center) and balanced it against the drums. The bass lacked definition, so I boosted +12 dB at 3 kHz and cut 3 dB at 250 Hz.

Then the guitars came in left and right; they balanced each other on the stereo stage.

I panned the lead vocal to center, and balanced it with the instruments. It needed no EQ except for +3 dB at 12 kHz.

Finally, I added the harmony vocals panned half-left and half-right, and I blended them with the lead vocal so that everyone could be heard equally, yet no one was louder than the lead vocalist. The harmony vocals required a little low-end rolloff to compensate for the proximity effect of the microphones, as well as a slight high-end rolloff to reduce sibilance.

With the balance well underway, I added a little reverb to everything but the bass and kick drum. The "small room" setting on the Alesis Microverb seemed to suit the music best because it was recorded in a small club.

The kick drum was on the same track as the rest of the drums. I wanted to add reverb to all the drums except the kick drum. To do this, I rolled off the low frequencies in the reverb-return signal. Since the kick drum is mainly a low-frequency signal, not much reverb was audible on the kick drum.

I practiced the mix several times by hitting the zero return button on the Fostex, playing the tape, and making console adjustments. I made a cue sheet that noted fader changes occurring at various tape-counter marks and practiced these changes. I set the recording levels to peak at +3 VU maximum and touched up EQ and reverb.

I was satisfied with the mix; it sounded solid, every instrument and vocal was clearly audible, and reverb enhanced the recording but was not excessive. The tonal balance was similar to that of good commercial recordings played through my headphones. Distortion was inaudible. The Fostex multitrack contributed very little noise; more was added by the Revox 2-track, but the noise level was still acceptable. By far the greatest source of noise was the rhythm-guitar amplifier.

Now I was ready to record the mix. I started the Revox A77 in record mode and hit play on the Fostex multitrack. Following the cue sheet, I made fader and EQ changes as needed. The mixdown for the first song was complete.

Next, I cued up the tape for the next song to be mixed, and repeated the entire process. Each song required a different designation strip. I noted the console settings for each song for future reference.

After all the mixes were recorded (20 songs), I labeled the master tapes. Then I edited out the pauses between songs, and stored the tapes tail-out to reduce print-through. I made a safety copy of the master tapes as a backup in case the masters were lost or destroyed.

Playback

Finally, I made a Dolby C cassette copy of the master tape to play for the band. I was with the band members when they played the tape, which was fortunate because their cassette deck and loudspeakers were inferior and needed a lot of tonal adjustment to reproduce the tonal balance I had heard over the headphones. If possible, always carry your own high-quality equipment with you for playback—don't rely on the speakers the band has available.

The band was delighted with the sound but unhappy with the performance. Consequently, the band couldn't use the tape as an audition demo, but the musicians learned from their mistakes—and called for another session. They found out which songs worked the best, and recorded only those songs for their final demo tape.

As we've seen, some home recording equipment can be used to make a professional-quality demo, as long as care is taken at every step.

11 Judging Sound Quality

Give a musician a recorder-mixer and ask him or her to do a mix. It sounds great. Then give another musician the same equipment and again ask for a mix. It sounds terrible. What happened?

The difference lies mainly in their ears—their critical listening ability. The first musician has a clear idea of what he or she wants to hear and how to get it. The second musician hasn't acquired the ability to recognize good sound.

That ability is essential. By knowing what to listen for, you can improve your artistic judgments during recording and mixdown. You'll be able to hear errors in microphone placement, equalization, and so on, and to correct them.

To train your hearing, try to separate recorded sound into its components—such as tonal balance, noise, reverberation—and concentrate on each one in turn. It's easier to hear sonic flaws if you are focusing on a single aspect of sound reproduction. This chapter is a guide to help you do this.

The precise translation of sound to tape is not always the goal in recording popular music. Although the aim may be to reproduce the original sound (as in an audition or live gig tape), you may also want to play with that sound to create a new sonic experience, or to do some of both. In fact, the artistic manipulation of sounds through studio techniques has become an end in itself. Apparently the philosophy is that creating an interesting new sound is as valid a goal as recreating the original sound. There are two games to play here, each with its own measures of success. If the aim of a recording is realism or accurate reproduction, the recording is successful when it sounds like the live instruments sounded.

But when the goal is to enhance the sound or produce special effects (as in most pop-music recordings), the aesthetic is less defined. The live sound of a pop group could be a reference, but pop-music recordings generally sound better than live performances. Recorded vocals are clearer and less harsh, the bass is cleaner and tighter, and so on. A standard for the quality of the sound of pop music reproduced over speakers has developed apart from accurate reproduction.

Good Sound in a Pop-Music Recording

Currently, a pop recording would sound good if it had these characteristics: good mix, wide frequency range, good tonal balance, cleanness, and clarity. Quality recordings also sound smooth and spacious and have presence, wide and detailed stereo imaging, and sharp transients. Dynamic range is wide but controlled, and special effects are creative and tasteful. Let's explore each of these qualities to know what to listen for. We'll assume the monitor system is accurate, so that any colorations heard are in the recording and not in the monitors.

Good Mix

In a good mix, the instruments and vocals are in a pleasing volume balance with each other. Everything can be clearly heard, yet nothing is obtrusive. The more important instruments or voices are loudest; less-important parts are in the background.

A successful mix goes unnoticed. When all the tracks are balanced correctly, nothing sticks out and nothing is hidden. Note that there's a wide latitude for musical interpretation and personal taste in making a mix.

Sometimes you don't want everything to be clearly heard. On rare occasions you may want to mix certain tracks in subtly for a subconscious effect.

The mix must be appropriate to the style of the music. For example, a mix that's right for rock music usually won't work for country music. A rock mix typically emphasizes the drums and has the vocals only slightly louder than the accompaniment. In contrast, a country mix emphasizes the vocals and the drums are used as "seasoning" in the background. This distinction is lessening as country music approaches a pop sound.

Level changes during the mix should be subtle, or else instruments will "jump out" for a solo and "fall back in" afterwards. Move your faders slowly or set them to preset positions during pauses in the music. Nothing in a mix sounds more amateurish than a solo that starts too quietly and then comes up as it plays: you can hear the engineer working the fader.

Wide Frequency Range

"Wide frequency range" means extended low-frequency and high-frequency response. Cymbals should sound crisp and distinct, but not sizzly or harsh; kick drum and bass should sound deep, but not overwhelming or muddy. A wide frequency range results from using high-quality microphones and recorders, good tape, high tape speed, and clean tape heads.

Good Tonal Balance

The overall tonal balance of a recording should be neither bassy nor trebly. That is, the perceived spectrum should not emphasize low frequencies or high frequencies. Low bass, mid-bass, midrange, upper midrange, and highs should be heard in equal proportions, as diagrammed in Figure 11-1. Emphasis of one frequency band over the others eventually causes listening fatigue.

Recorded tonal balance is inversely related to the frequency response of the studio's monitor system. If the monitors have an extreme high-frequency rolloff, the engineer will compensate by boosting highs in the recording to make the monitors sound right. The result is a bright recording.

Before doing a mix, it helps to play over the monitors some recordings whose sound you admire. This helps you become accus-

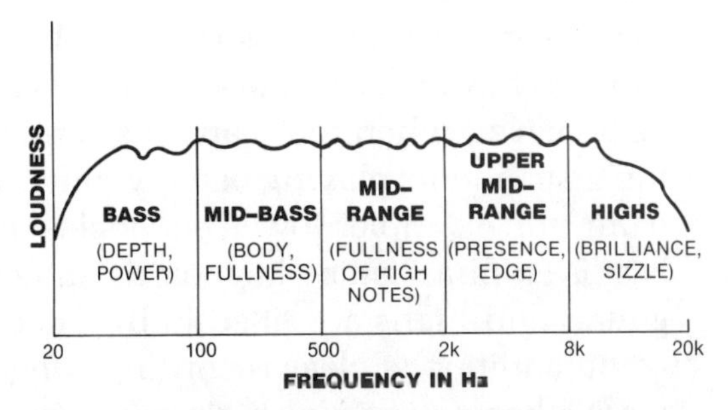

Figure 11-1. Loudness vs. frequency of a good-sounding pop recording.

tomed to a commercial spectral balance. After your mix is recorded, play it back, switching between your mix and a commercial recording. This comparison will indicate how well you matched a commercial spectral balance. Of course, you may not care to duplicate what others are doing.

In pop-music recordings, the tonal balance or timbre of individual instruments does not have to be natural. Still, many listeners want to hear a realistic timbre from acoustic instruments such as guitar, flute, sax, or piano. The reproduced timbre depends on the musical instruments themselves and on microphone frequency response, microphone placement, and equalization.

Cleanness

A "clean" recording is free of noise and distortion. Tape hiss, hum, and distortion are inaudible in a good recording. "Distortion" in this case means that which is added by the recording process, not the distortion which might already be present in the sound of electric-guitar amps or Leslie speakers.

"Clean" also means "not muddy"—free of low-frequency room reverberation and leakage.

A "clean mix" is one that is uncluttered, free of excess instrumentation. This is achieved by arranging the music so that similar parts don't overlap and too many instruments don't play at once. Usually, the fewer the instruments, the clearer the sound. Too many overdubs can muddy the mix.

Clarity

In a "clear" recording, instruments do not crowd, or mask, each other's sound. They are separate and distinct and they blend well. Clarity can be achieved when too many instruments don't play at once or when the instruments playing occupy different areas of the frequency spectrum (for example, low frequencies are provided by the bass, mid-bass is emphasized by keyboards, upper midrange is provided by lead guitar, and highs are filled in by the cymbals).

In addition, a clear recording adequately reproduces each instrument's harmonics (the high-frequency response is not rolled off).

Smoothness

Now we get into some subtler aspects of sound. "Smooth" means uncolored, easy on the ears, not harsh. Sibilants are clear but not piercing. A smooth, effortless sound allows relaxation; a strained or irritating sound causes tension in the listener.

Smoothness is a lack of sharp peaks or dips in the frequency response as well as a lack of excessive boost in the midrange or upper midrange.

Presence

This is the apparent sense of closeness of the instruments—a feeling that they are present in the listening room. Synonyms include clarity, detail, punch. Presence is achieved by close miking, overdubbing, and using microphones with a presence peak or emphasis around 5 kHz. Upper-midrange boost helps too. Most instruments have a frequency range that, if boosted, makes the instrument stand out more clearly or become better defined.

Note that presence sometimes conflicts with smoothness. That's because presence often involves an upper-midrange boost, while a smooth sound is free of such emphasis. You'll have to find a tasteful compromise between the two.

Spaciousness

"Spacious," or "airy," means having a sense of air around the instruments. Without air, or ambience, instruments sound like they're isolated in stuffed closets. Spaciousness is achieved by adding artificial reverberation to the recording. The amount to use varies from almost none to a lot, depending on the song and musical style.

Sharp Transients

The attack of cymbals and drums generally should be sharp and clear. Bass guitar and piano may or may not require sharp attacks, depending on the song.

Tight Bass and Drums

The kick drum and bass guitar should "lock" together so that they sound like a single instrument: a bass with a percussive attack. The drummer and bassist should work out their parts together to hit accents simultaneously.

To further tighten the sound, the kick drum is damped, the bass is recorded direct, and they are both equalized for presence and clarity.

Good Stereo Imaging

Stereo means more than just "left" and "right." Usually, tracks should be panned to many points across the "stereo stage" between the playback loudspeakers. Some instruments should be hard left or hard right; some should be in the center; others should be half-left or half-right. Try to achieve a stereo stage that is well balanced between left and right, as shown in Figure 11-2. Instruments occupying the same frequency range should be panned to opposite sides of center.

You may want some tracks to be unlocalized. Backup choruses and strings should be spread out rather than appearing as point sources. Stereo keyboard sounds can wander between speakers. A lead-guitar solo can have a fat, spacious sound.

There should be some sonic depth. Some instruments should sound close or up front, while others should sound farther away.

If you intend your stereo imaging to be realistic (say, for a jazz combo), then the reproduced ensemble should simulate the spatial layout of the live ensemble. If you're sitting in an audience listening to a jazz quartet, you might see and hear drums on the left, piano on the right, bass in the middle, and sax slightly right. The drums and piano are not point sources, but are somewhat spread out. If spatial realism is the goal, you should hear the same ensemble layout between

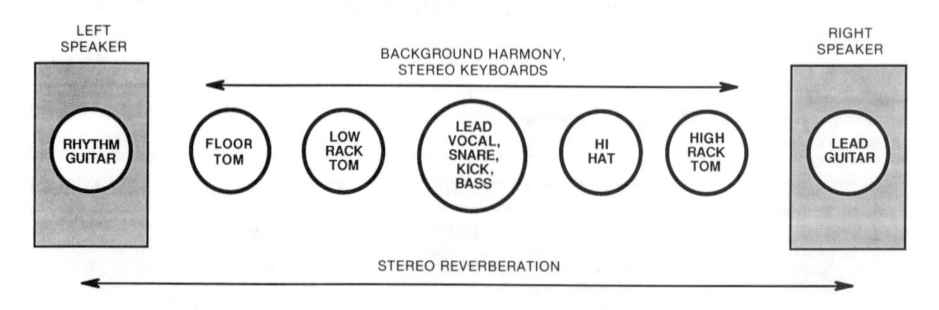

Figure 11-2. Example of image placement between speakers.

your speakers. Often the piano and drums are spread all the way between speakers—an interesting effect, but unrealistic.

Pan-potted mono tracks often sound artificial, in that each instrument sounds isolated in its own little space. Some stereo reverberation surrounding the instruments helps to "glue" them together.

Wide but Controlled Dynamic Range

Dynamic range is the range of volume levels from softest to loudest. A recording with a wide dynamic range becomes noticeably louder and softer, adding excitement to the music. This is achieved by avoiding excessive compression (automatic volume control). An overly compressed recording sounds "squashed": crescendos and quiet interludes lose their impact.

Some compression or gain-riding is needed for vocals because the vocalist's dynamic range exceeds that of the instrumental backup. A compressor can even out extreme level variations, keeping the vocals relatively constant. Bass guitar also often needs compression for the same reason.

Interesting Sounds

The recorded sound may be too flat—lacking character or color. In contrast, a recording with creative production has interesting instrumental sounds and typically uses equalization and special effects. Some of these effects are echo, reverberation, doubling, chorus, flanging, compression, and stereo effects. Making sounds colorful can conflict with accuracy or fidelity, so effects and equalization should be used with discretion.

Suitable Production

Marshall McLuhan's observation that "the medium is the message" could be considered a general rule of recording aesthetics. The way a recording sounds should imply the same message as the musical style or lyrics. In other words, the sound should be appropriate for the tune being recorded. For example, some rock music is rough and raw; the sound should be, too. A clean, polished production doesn't always work for high-energy rock and roll. There might even be a lot of leakage or ambience in order to suggest a garage studio or nightclub environment. The role of the drums is important, so they should be loud in the mix and toms should ring.

New Age, Disco, and Middle-of-the-Road musical styles are slickly produced. The sound is usually tight, smooth, and spacious.

Country music is about stories or feelings, so the bass and vocals are emphasized for warmth and emotion. Acoustic guitars and drums are miked at a respectful distance, giving an airy, natural effect. Blues and folk music are usually recorded with little or no production EQ or effects.

Actually, no style of music is locked into a particular style of production. You tailor the sound to complement the music of each individual tune. Doing this may break some of the "rules" of good sound, but that's usually okay as long as the song is enhanced by its sonic presentation.

Training Your Hearing

We've covered a long list of things to listen for. The critical process is easier if you focus on one aspect of sound reproduction at a time. You might concentrate first on the tonal balance—try to pinpoint what frequency ranges are being emphasized or slighted. Next listen to the mix, the clarity, and so on. Soon you'll have a detailed appreciation of the sound quality of your recording.

Here is a checklist you can use to evaluate the sound quality of a recording.

Well mixed?	Nothing too loud or too quiet
Wide range?	Extended lows and highs
Tonally balanced?	No frequency range too loud or quiet
Clean?	Free of noise and distortion
Clean?	Free of low-frequency room reverberation and leakage
Clean?	Uncluttered mix
Clear?	Instruments distinct
Smooth?	Not harsh—no excessive upper midrange
Presence?	Clarity, detail, punch, closeness
Spacious?	Has reverberation or ambience

Sharp transients?	Clear percussive attacks
Tight bass and drums?	Synchronous playing, damped kick
Good stereo?	Well balanced, various image locations
Good stereo?	Instruments not too isolated, sense of depth
Wide but controlled dynamic range?	Vocals not too loud or too quiet
Interesting sounds?	Effects, production tricks
Suitable production?	Production suits the musical style

Developing an analytical ear is a continuous learning process. Train your hearing by listening carefully to recordings—both good and bad. Make a checklist of all the qualities mentioned in this chapter. Compare your own recordings to live instruments, and to commercial recordings, to see what you're doing right or wrong.

A pop-music record that excels in all the attributes of good sound is "The Sheffield Track Record" (Sheffield Labs, Lab 20), engineered and produced by Bill Schnee. In effect, it's a course in state-of-the-art sound, and it should be required listening for any recording engineer or producer.

Another record with brilliant production is "The Nightfly" by Donald Fagen (Warner Brothers 23696-1); engineered by Roger Nicols, Daniel Lazarus, and Elliot Scheiner; produced by Gary Katz; and mastered by Bob Ludwig. The sound is razor sharp, elegant, and tasteful, and the music just pops out of the speakers.

Here are four more examples of outstanding rock production. They set high standards to work toward.

Song: "I Need Somebody"
Artist: Bryan Adams
Producer: Bob Clearmountain

Song: "The Power of Love"
Artist: Huey Lewis & The News
Producer: Huey Lewis & The News

Album: "Synchronicity"
Artist: The Police
Producer: Hugh Padgham and The Police

Album:	"Thriller"
Artist:	Michael Jackson
Engineer:	Bruce Swedien
Producer:	Quincy Jones

Once you're making recordings that are technically competent—clean, natural, and well mixed—the next stage is to produce imaginative sounds. You're in command; you can tailor the mix to sound any way that pleases you or the people you're recording.

The supreme achievement, finally, is to produce recordings that sound beautiful.

Troubleshooting Bad Sound

Now you know how to recognize good sound, but can you recognize bad sound? Suppose you're monitoring a recording in progress, or listening to a recording you've already made, and something doesn't sound right. How can you pinpoint what's wrong and fix it?

The remainder of this chapter includes step-by-step procedures to solve common audio problems (equipment maintenance is not covered). Read down the "bad sound" descriptions until you find one matching what you hear. Then try the solutions until your problem disappears.

This troubleshooting guide is divided into three main sections:

1. bad sound on all recordings, including those from other studios
2. bad sound on tape playback only; mixer output sounds are okay
3. bad sound from your mixer

Before you start, check for faulty cables and connectors. Also check all control positions; rotate knobs and flip switches to clean the contacts.

Bad Sound on All Recordings

Upgrade your monitor system. Adjust tweeter and midrange controls on speakers, relocate speakers, improve room acoustics, equalize the monitor system, try different speakers, upgrade the power amp and speaker cables.

Bad Sound Only on Tape Playback

1. Dull sound or dropouts
   A. Check that the oxide side of the tape is against the heads.
   B. Clean and demagnetize the tape path.
   C. Try another brand of tape.
   D. Align the tape heads. Calibrate the electronics.
   E. Do maintenance on the tape transport.
   F. Check and replace the tape heads if necessary.

2. Distortion
   A. Reduce the recording level.
   B. Increase the bias level.

3. Tape hiss
   A. Increase the recording level.
   B. Use a type of noise reduction, such as Dolby or dbx.
   C. Use better tape.
   D. Align the tape recorder.

Bad Sound from Your Mixer Output

1. Muddy (excessive leakage)
   A. Place the microphones closer to the sound sources.
   B. Spread the instruments farther apart to reduce the level of the leakage.
   C. Place the instruments closer together to reduce the delay of the leakage.
   D. Use directional microphones (such as cardioids).
   E. Overdub the instruments.
   F. Record the electric instruments direct.
   G. Use baffles (goboes) between the instruments.
   H. Deaden the room acoustics (add absorptive material, flexible panels, or slot absorbers).

I. Filter out frequencies above and below the spectral range of each instrument.

J. Turn down the bass amp in the studio.

2. Muddy (excessive reverberation)

A. Reduce the aux-send levels or the aux-return levels.

B. Place the microphones closer to the sound sources.

C. Use directional microphones (such as cardioids).

D. Deaden the room acoustics.

E. Filter out frequencies below the fundamental frequency of each instrument.

3. Muddy (lacking highs, dull or muffled sound, poor transient response)

A. Use microphones with better high-frequency response, or use condenser mics instead of dynamics.

B. Change the microphone placement. Put the microphone in a spot where there are sufficient high frequencies. Keep the high-frequency sources, such as cymbals, on-axis to the microphones.

C. Use small-diameter microphones, which generally have a flatter response off-axis.

D. Boost the high-frequency equalization.

E. Change musical instruments, replace guitar strings, or replace drum heads.

F. When bouncing tracks, record bright-sounding instruments last to reduce generation loss.

G. Use a better tape.

H. Avoid excessive recording levels with bright-sounding instruments, because the recorder's high-frequency response gradually rolls off as the recording level is increased. This is especially true of cassette recorders.

I. Use exciter signal processors, such as the Aphex Aural Exciter.

J. Use a direct box on the electric bass. Have the bassist play percussively or use a pick. When compressing the bass, use a long attack time to allow the note's attack to come through. Note: Some songs don't require sharp bass attacks—just do whatever's right for the song.

K. Damp the kick drum with a pillow or blanket, and mike it next to the center of the head near the beater.

4. Muddy (lacking clarity)

A. Use fewer instruments in the musical arrangement.

B. Equalize instruments differently so that their spectra don't overlap.

C. Try less reverberation.

D. Delay the reverberation-send signal by about 30 to 70 ms.

E. Using equalizers, boost the presence range of the instruments that lack clarity.

5. Distortion

A. Switch in the pad built into the microphone (if any).

B. Increase the input attenuation (reduce the input gain), or plug in a pad between the microphone and mic input.

C. Readjust gain-staging: set the faders and pots to their design centers (shaded areas).

6. Bad tonal balance (nasal, rough, boomy, dull, shrill, etc.)

A. Change musical instruments, change guitar strings, change reeds, etc.

B. Change the microphone placement. If the sound is too bassy with a directional microphone, you may be getting proximity effect. Mike farther away or roll off the excess bass.

C. Use the 3:1 rule of microphone placement to avoid phase cancellations. When multiple microphones are mixed to the same channel, the distance between microphones should be at least 3 times the mic-to-source distance.

D. Try another microphone. If the proximity effect of a cardioid mic is causing a bass boost, try an omnidirectional mic instead.

E. If you must place a microphone near a hard reflective surface, try a boundary microphone on the surface to prevent phase cancellations.

F. Change the equalization. Avoid excessive boost.

G. Use equalizers with a broad bandwidth, rather than a narrow, peaked response.

8. Lifeless sound (unexciting)

A. Work on the live sound of the instruments to come up with unique effects.

B. Add equalization or special effects (reverberation, echo, doubling, etc.)

C. Use and combine recording equipment in unusual ways.

D. Try overdubbing little vocal licks or synthesized sound effects.

9. Lifeless sound (dry or dead acoustics)

A. If leakage is not a problem, put the microphones far enough from instruments to pick up wall reflections. If you don't like the sound this produces, try the next suggestion.

B. Add artificial reverberation or echo to dry tracks. Note: Not all tracks require reverberation, and some songs may need only a little reverberation in order to sound intimate.

C. Use omnidirectional microphones.

D. Add hard reflective surfaces in the studio, or record in a hard-walled room.

E. Allow a little leakage between microphones. Put microphones far enough from instruments to pick up off-mic sounds from other instruments. Don't overdo it, though, or the sound will become muddy and track separation will become poor.

9. Noise (hiss)

A. Check for noisy guitar amps or keyboards.

B. Switch out the pad built into the microphone (if any).

C. Reduce console input attenuation (increase input gain).

D. Use a more sensitive microphone.

E. Use a quieter microphone (one with low self-noise).

F. Increase the sound-pressure level at the microphone by miking closer. If you're using PZMs, mount them on a large surface or in a corner.

G. Use a noise gate.
H. Use a lowpass (high-cut) filter.

10. Noise (low-frequency rumble)
   A. Reduce air-conditioning noise or temporarily shut off the air conditioning.
   B. Use a highpass (low-cut) filter set around 40 to 80 Hz.
   C. Use microphones with limited low-frequency response.

11. Noise (thumps)
   A. Change the microphone position.
   B. Change the musical instrument.
   C. Use a highpass filter set around 40 to 80 Hz.
   D. If the cause is mechanical vibration traveling up the microphone stand, put the microphone in a shock-mount stand adapter. Or use a microphone that is less susceptible to mechanical vibration, such as an omnidirectional microphone, or a unidirectional microphone with a good internal shock mount.
   E. Use a microphone with a limited low-frequency response.

12. Hum
   This is a subject in itself. See Chapter 3 for causes and cures of hum, as well as the Electric-Guitar section in Chapter 7.

13. Pop (explosive breath sounds in the vocalist's microphone)
   A. Place the microphone above or to the side of the mouth.
   B. Place a foam windscreen (pop filter) on the microphone.
   C. Place the microphone farther from the vocalist.
   D. Use a microphone with a built-in pop filter (ball grille).
   E. Use an omnidirectional microphone, because it is likely to pop less than a directional (cardioid) microphone.

14. Sibilance
   A. Use a de-esser signal processor.
   B. Place the microphone farther from the vocalist.
   C. Place the microphone toward one side of the vocalist, rather than directly in front.

D. Cut equalization in the range from 5 to 10 kHz.

E. Change to a duller-sounding microphone.

15. Bad mix

A. Change the mix.

B. Compress vocals or instruments that occasionally get buried.

C. Change the equalization on instruments you want to stand out.

D. During mixdown, continuously change the mix to highlight certain instruments according to the demands of the music.

16. Unnatural dynamics

A. Check the tracking of noise-reduction units. For example, a 10-dB level increase at the input of the encode unit should appear as a 10-dB level increase at the output of the decode unit.

B. Use less compression.

C. Avoid overall compression.

17. Instruments sound too isolated, or sound like they're in different acoustic environments

A. In general, allow a little crosstalk between the left and right channels. If tracks are totally isolated, it's hard to achieve the illusion that all the instruments are playing in the same room at the same time. You need some crosstalk or correlation between channels. Some right-channel information should leak into the left channel, and vice-versa.

B. Place the microphones farther from the sound sources to increase leakage.

C. Use omnidirectional microphones to increase leakage.

D. Use stereo reverberation or echo.

E. Pan the echo returns or reverberation returns to the channel opposite the channel of the dry sound source.

F. Pan the extreme left-and-right tracks slightly toward the center.

G. Make the aux-send levels more similar for various tracks.

H. To give a lead-guitar solo a fat, spacious sound, use a stereo chorus. Or send its signal through a delay unit, pan the direct sound hard left and pan the delayed sound hard right.

18. The mix lacks depth

A. Achieve depth by miking instruments at different distances.

B. Use varied amounts of reverberation on each instrument. The higher the ratio of reverberant sound to direct sound, the more distant the track sounds.

C. If your mixing board has a pre-post switch by the aux-send control, set the switch to "pre" to move an instrument's sound closer or farther as you move the fader.

Conclusion

This chapter described a set of standards for good sound quality. These standards are somewhat arbitrary, but there's a need for some guidelines to go by in judging the effectiveness of the recording.

We also discussed causes and cures of various audio ailments. The next time you hear something you don't like in a recording, you'll know better how to define the problem and how to fix it.

12 MIDI Studio Recording Procedures

You might record your demo with a home MIDI (Musical Instrument Digital Interface) studio including one or more synthesizers, sampling keyboards, a drum machine, and a sequencer. Other texts explain this equipment in detail, so we'll give just brief definitions here.

MIDI Studio Uses

MIDI is a standard connection between electronic musical instruments and computers that allows them to communicate with each other. The MIDI signal is a digital bitstream—not an audio signal—that runs at 31,250 bits per second, carrying copies of gestures you make on a keyboard or drum pads. Up to 16 channels of information can be sent on a single MIDI cable.

With MIDI, you can:

1. combine the sounds of two electronic musical instruments by playing them both with one keyboard.
2. create the effect of a band playing. To do this, you record keyboard performances into a computer memory, edit the recording, and have the recording play through several synthesizers and a drum machine in sync.
3. automate a mixdown or effects.
4. create the sound of any instrument with an electric guitar or breath controller.

For more information on MIDI, see Appendix C, Reference Sources.

MIDI Studio Equipment

A MIDI studio might include some or all of the following equipment.

- Synthesizers: keyboard instruments that create sounds electronically with oscillators. A *multitimbral synthesizer* can play two or more patches at once. A patch is a sound preset (an instrumental timbre), such as a synthesized piano, bass, or snare drum. A *polyphonic synthesizer* can play several notes at once (chords) with a single patch. A *sound generator* is a synthesizer without a keyboard.
- Sampler: a device (such as the one shown in Figure 12-1) that records short sound events, or samples, in computer memory. A sample is a digital recording of one note of a real sound source, such as a flute note, a bass pluck, or a drum hit.

  Often a sampler is built into a sample-playing keyboard, which resembles an electronic piano. It contains samples of several different instruments. When you play on the keyboard, the sample notes are heard. The higher the key you press, the higher the pitch of the reproduced sample.
- Drum machine: a device that plays samples of various percussion instruments, including a drum set. It also records and plays back drum patterns that you play or program with built-in keys or drum pads.
- Sequencer: a device that records performance gestures, such as keypresses on a piano-style keyboard, into computer memory. Unlike a tape recorder, a sequencer does not record audio. Instead, it records note parameters: the key number, note-on and note-off signals, and so on.

  The sequencer can be a stand-alone unit (such as that shown in Figure 12-2), a circuit build into a keyboard instrument, or a computer running a sequencer program. Like a multitrack tape recorder, a sequencer can record eight or more tracks, with each track containing a performance of a different instrument.

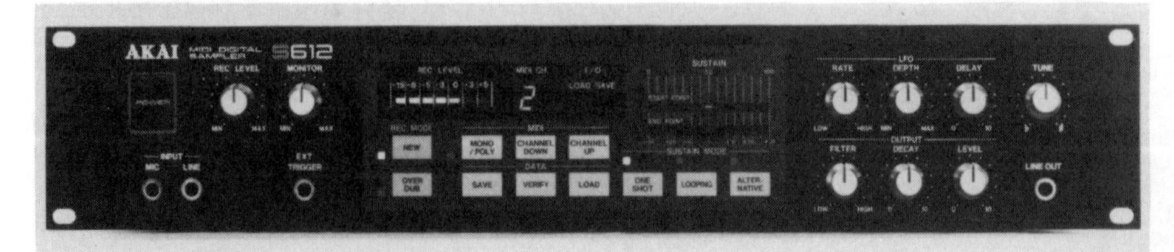

Figure 12-1. Akai S612 MIDI Digital Sampler. *(Courtesy of Akai Corporation)*

Figure 12-2. The Alesis MMT-8 Multitrack MIDI Recorder. *(Courtesy of Alesis Corporation)*

- Amplifier and speakers, or powered speakers: monitoring systems which let you hear what you're performing and recording. These are usually small speakers set up in a close-field arrangement (about 3′ apart and 3′ from you).
- Personal computer system (optional): a computer, disk drive, monitor, and perhaps a printer. Generally, the computer is used to run a sequencer program, which replaces a sequencer. Compared to a stand-alone sequencer, which has an LCD screen, the computer monitor screen displays much more information at a glance, making editing easier and more intuitive (see Figure 12-3).

  The computer also runs other useful programs. A *voice editor* program lets you manipulate the parameters that make up a MIDI instrument's patches or sounds. A *sample editor* displays a sample's waveform on your computer screen and lets you modify it. A *librarian* enables you to transfer patches between MIDI instruments and your computer, rename or rearrange the patches, and store them to disk. A *notation program* converts your sequenced performance to standard musical notation and prints it.

  Some popular computers for music composing are the Commodore 64 and 128; Atari ST; Macintosh; IBM PC; and the

Yamaha C1 Music Computer, an IBM-compatible portable unit with MIDI and SMPTE connectors.

- MIDI-computer interface (optional): a device that plugs into a user port in your computer and converts MIDI signals into computer signals and vice versa. You need this only if you're using a computer in your system.
- Recorder-mixer (optional): a combination mixer and cassette recorder for recording vocals and acoustic instruments. A more elaborate studio might use a separate mixer and 8-track recorder (either open-reel or cassette).
- Tape synchronizer (optional): a device, such as the one shown in Figure 12-4, that synchronizes tape tracks with sequencer tracks and makes the sequencer start at the same place in the song that you start the tape.
- Mixer (optional): a device that blends the audio outputs of two or more synthesizers, or a synth and a drum machine, into a single stereo signal.
- 2-track recorder: for use in recording the stereo mix of all your sound sources. The tape made on this recorder is the final product. The recorder can be open-reel, cassette, DAT, or a VCR with a digital audio adapter.

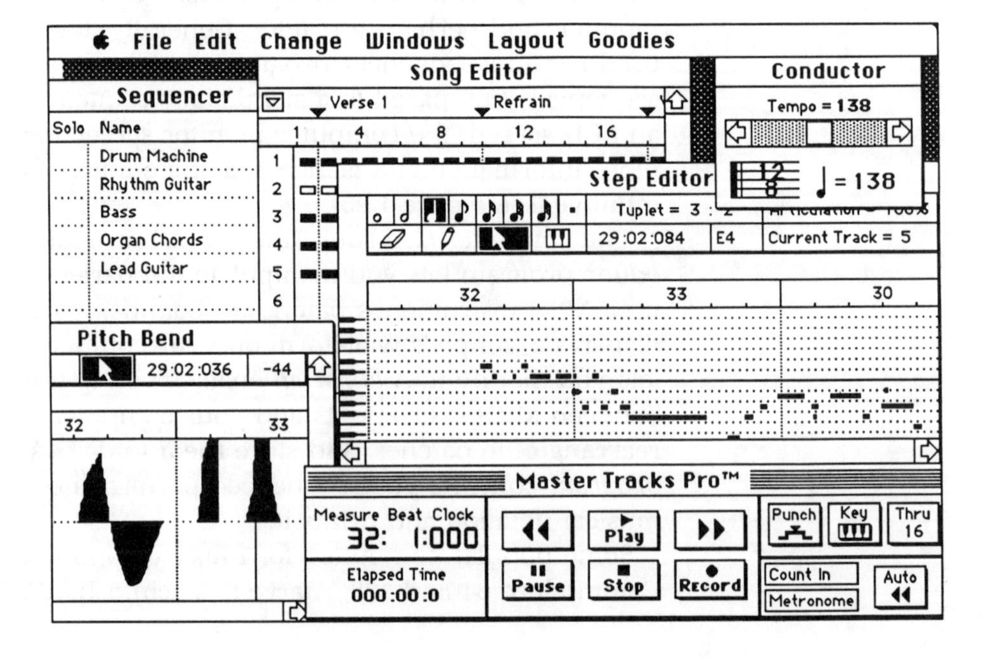

Figure 12-3. Monitor-screen shot of Passport Master Tracks Pro MIDI software. *(Courtesy of Passport Designs, Inc.)*

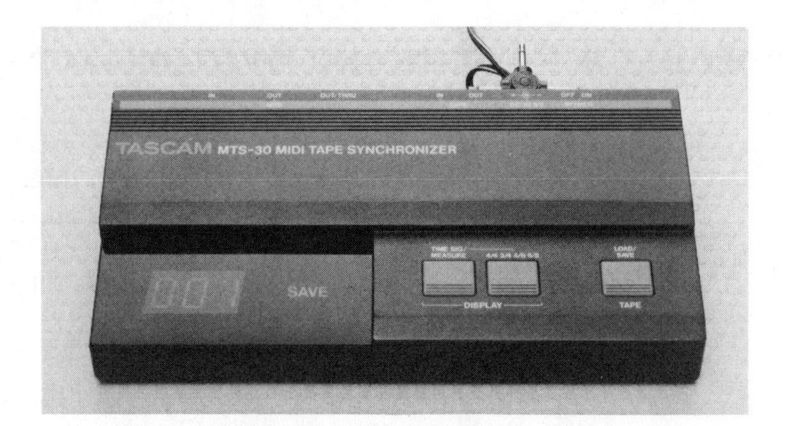

Figure 12-4. Tascam MTS-30 MIDI Tape Synchronizer. *(Photo courtesy of TEAC America, Inc./TASCAM Professional Division of TEAC)*

- Audio cables: which carry audio signals, connecting (via ¼″ phone plugs) synthesizers and drum machines with your mixer line-input connectors.
- MIDI cables: which carry MIDI signals and are used to connect (via 5-pin DIN plugs) synths, drum machines, and computers so that they can communicate with each other.
- Power outlet strip: a row of electrical outlets to power all your equipment. It's a good idea to have surge protection in the strip.
- Equipment stand: a system of tubes, rods, and platforms that supports all your equipment in a convenient arrangement. It provides user comfort, enables use of shorter cable lengths, and saves floor area for other activities. Two manufacturers of MIDI studio stands are Ultimate Support Systems and Invisible Stands.
- Workstation: a system comprising the equipment above is called a musical workstation. The components might be separate or combined in a single package. For example, a workstation might include a keyboard with a built-in synthesizer, sampler, and sequencer—everything you need (except a monitor speaker system) to compose, perform, and record instrumental music. If you want to record songs with vocals, you also need a tape recorder or recorder-mixer. Some workstations include drum sounds, so you can get by without a separate drum machine. Two examples of workstations are the Roland W-30 and the Korg M1.

The remainder of this chapter describes recording procedures for several different MIDI studio setups, from simple to complex. You'll better understand the procedures for the complex systems if you read about the simple systems first.

Recording a Polyphonic Synthesizer

This is the simplest method of recording. The setup is shown in Figure 12-5. A synthesizer is connected to a MIDI interface that is connected to a computer running a sequencer program. You'll play chords and melody, record them with the sequencer, and play back the sequence through your synthesizer. The basic steps follow:

1. Set your sequencer in record mode.
2. Play a tune on your keyboard.
3. Play back the sequencer recording to hear it. Your performance will be duplicated with the synthesizer.
4. Correct mistakes by punching in.
5. Arrange the song by combining various sequences.
6. When the sequenced performance is correct and complete, start the sequencer and record the synth output. That recording is the final product.

Let's look at each step in more detail.

1. Set Your Sequencer in Record Mode

Choose a tempo in your sequencer and choose the track you want to record on. On your computer keyboard, push the record key (designated in the sequencer instructions). You'll hear a metronome ticking at the tempo you set.

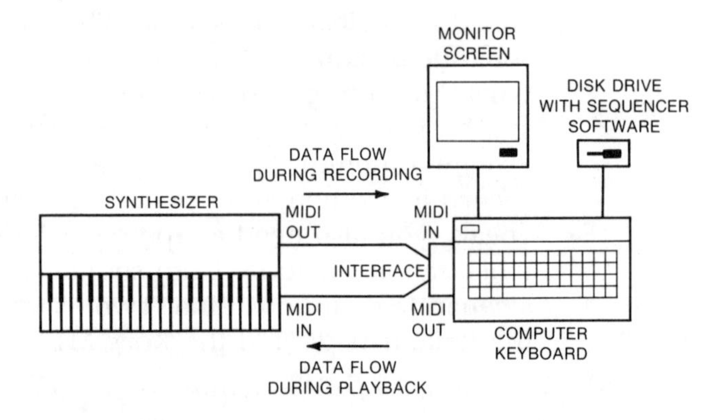

Figure 12-5. A synthesizer connected via a MIDI interface to a computer running a sequencer program.

2. Play Music on Your Keyboard

Listen to the sequencer's metronome and play along with its beat. The sequencer will keep track of the measures, beats, and pulses. When you press the stop key on your computer keyboard, the sequencer will stop recording and will go to the beginning of the sequence (the top of the tune).

Another way to record your performance is in step-time, one note at a time. If the part is difficult to play rapidly, you can also set the sequencer tempo very slow, record while playing the synth at that slow tempo, then play back the sequence at a faster tempo.

3. Play Back the Sequencer Recording

On your computer keyboard hit the play key (designated in the sequencer-program instructions). You'll hear the sequence playing through your synthesizer.

4. Punch In-Out to Correct Mistakes

You can correct mistakes by punching into record mode before the mistake, recording a new performance, and then punching out of record mode. Here's one way to do it:

1. Go to a point in the song a few bars before the mistake.
2. Just before you get to the mistake, punch into record mode and play a new, correct performance.
3. As soon as you finish the correction, punch out of record mode.

Alternatively, you can use *Autopunch.* With this feature, the computer automatically punches in and out at preset times; all you have to do is play the corrected musical part. Perform an autopunch as follows:

1. Using the computer keyboard, set the punch-out point (the measure, beat, and pulse where you want to go out of record mode).
2. Set the punch-in point (just before the part you want to correct).
3. Set the cue point (where you want the track to start playing before the punch).
4. Hit the play key on your computer.
5. When the screen indicates punch-in mode, play the corrected part.

6. The sequencer will automatically punch out at the specified point in the song.

These punch-in routines were done in real time. You can also punch in-out in step time:

1. Go to a point in the song just before the mistake.
2. Set the sequencer to step-time mode.
3. Step through the sequence pulse by pulse, and punch into record mode at the proper point.
4. Record the proper note in step time.
5. Punch out of record mode.

5. Create a Song by Combining Sequences

Now your sequenced performance is perfect, so you can put together your composition. Many songs have repeated sections: the verse and chorus are each repeated several times. If you wish, you can play the verse and chorus once each, and save each as a separate sequence, which you can copy for all the places each occurs in the song.

You can rearrange song sections, and append one section to another, by pressing a few keys on the computer. You also can have any section replayed wherever desired in the song. Thus, you might build a song by having the computer play sections A, A, B, A, C, B, A.

To add variety to the song, you might want to have the synth play different patches at different parts of the song. For example, play a piano on the first verse, organ on the second, and marimba on the chorus. You record these program changes by putting your sequencer in record mode on another track and entering the appropriate program numbers at the right time on your synthesizer. Putting the program changes on a separate track makes it easy to edit them. You can punch in new program changes just as you can punch in new performances.

6. Play the Composition and Set Recording Levels

Plug your synthesizer's audio output (mono or stereo) into the line inputs of your 2-track recorder. Hit the play key on your computer keyboard, and set the recording level for your recorder:

Cassette: 0 maximum

Open-reel: +3 VU maximum

DAT: −3 dB maximum

7. Record the Synth Output

Once your levels are set, put your recorder in record mode and start the sequencer. This produces a finished product: a stereo recording of your song.

Recording a Drum Machine and a Synthesizer

Let's add a drum machine to the previous setup. Figure 12-6 shows how to connect the cables.

Suppose you've recorded a drum pattern using the drum machine's internal sequencer (it's easier to do this than to record the pattern on several tracks of an external sequencer). And you've recorded a synthesizer melody with an external sequencer. How do you synchronize the drum patterns in the drum machine with the synthesizer melody in the sequencer? In other words, how do you get them to play in sync, when both have different patterns recorded in different memories?

To synchronize the machines, you use a single MIDI clock (timing reference) that sets a common tempo for all the equipment. The MIDI clock is a series of bytes in the MIDI data stream that conveys timing information. The clock is like a conductor's baton movements, which keep all the performers in sync at the same tempo. The clock bytes are added to the MIDI performance information in the MIDI signal. The clock signal is 24, 48, or 96 pulses per quarter note (PPQ). That is, for every quarter note of the performance, 24 or more clock pulses (bytes) are sent in the MIDI data stream.

A sequencer or a drum machine can do the following with the clock pulses:

- send clock pulses from its MIDI-out connector

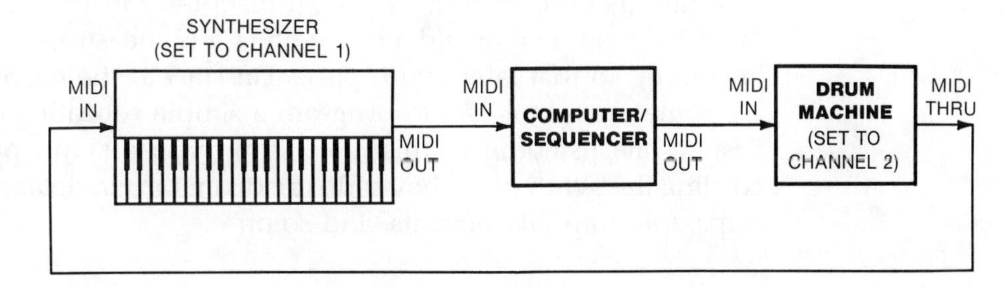

Figure 12-6. Equipment connections to make a sequencer drive a drum machine and a synthesizer.

- receive clock pulses at its MIDI-in connector
- echo (duplicate) incoming clock pulses through the MIDI-thru connector.

If a drum machine sends clock pulses, it sets the tempo and other devices follow it. If a drum machine receives clock pulses from a sequencer, the sequencer sets the tempo and the drum machine follows it.

To send clock pulses from a sequencer or drum machine, you set either device to "Internal clock" mode. To receive clock pulses, set the device to "External clock" or "MIDI clock" mode. Transmitting a clock pulse through the MIDI-thru connector is automatic. If your drum machine has only a MIDI-out connector, you enable "Echo MIDI in" so that the incoming pulses are echoed or repeated at the MIDI-out connector.

In the setup shown in Figure 12-6, the sequencer's clock drives both the drum machine and the synthesizer. The drum machine's internally recorded patterns play in sync with the synthesizer's sequencer-recorded melody. That procedure follows.

1. Record a Drum Pattern

The drum machine has a built-in sequencer that records what you tap on its pads. For example, you play a four-bar riff on the hi- hat and kick-drum keys. The riff is stored in memory. When you play it back, it repeats over and over (loops) every four measures. While this is happening, you can add a snare-drum back beat, and that combination will be stored. You can keep adding and storing the new combinations. You can mix the recording by adjusting the faders on the drum machine for each instrument. You store the finished four-bar riff as Pattern 1.

Next, you repeat the process for a different rhythmic pattern—say, a drum fill—and store this as Pattern 2. Then develop other patterns and store them. Finally, you make a song by chaining all the patterns together, as described in the drum machine's instruction manual.

It's a good idea to add a count-off (a few measures of clicks) at the beginning so that later synth parts can start at the correct time.

Some musicians like to program a simple repeating drum groove first. While listening to this, they improvise a synth part. After recording the synth part, they redo the drum part in detail, adding hand claps, tom-tom fills, accents, and so on.

2. Record a Synth Melody or Chords with an External Computer or Sequencer

3. Set the Drum Machine to Receive an External-Clock or MIDI-Clock Signal.

4. Set the Sequencer to Internal Clock and MIDI Drum.

5. Set MIDI Channels

Set the drum machine to channel 1. Set the sequencer synth track and the synthesizer to channel 2. In this way, the sequencer's recorded performance will play only the synthesizer. The MIDI clock will still control both devices, even though they are set to different channels.

6. Press the Play Key on Either the Computer or Sequencer

As the sequencer plays its recorded synth melody, the sequencer's clock pulses drive the drum machine and synthesizer at the same tempo. The drum machine plays its internally recorded patterns, while the synth plays the sequencer track.

Another way to synchronize a drum machine and a synthesizer is to record the drum patterns on one track of your sequencer. The advantage in this is that whenever you rearrange parts of the music in the sequencer, you also rearrange the drum part. So you don't have to change drum patterns each time you repeat or delete a verse or a chorus. Follow this procedure to record the drum patterns into your sequencer:

1. Record a drum pattern with the drum machine's internal sequencer.
2. Enable the drum machine's clock out and MIDI data out.
3. On your sequencer, turn off the MIDI-thru feature (if it has one).
4. Set the sequencer to external clock or MIDI clock mode and record track 1.
5. Hit the play key on the drum machine. The sequencer will record the drum pattern on track 1.

To play back the drum patterns you just recorded, follow this procedure:

1. Set the drum machine to External MIDI-clock mode.
2. Set the sequencer to Internal clock and MIDI drum.
3. Set track 1 and the drum machine to the same MIDI channel.
4. Load an empty pattern into the drum machine, so that the machine plays only the sequencer track.
5. Put the sequencer in play mode. The drum machine will play its sequencer track at the sequencer's tempo, and other synths connected to the sequencer will play their tracks on their channels.

Recording a Single Multitimbral Synthesizer

With this method, you play the parts for several different instruments (patches) on the same multitimbral keyboard and record each performance on a separate track of your sequencer. In playback, the sequence plays all the instruments in your synth and sounds like a band playing. Record the synth's output to get the final product. Here's a suggested procedure:

1. Select Patches

Select the patches or samples to be used in the composition. Often you don't know the patches in advance, so you choose them as you go. By punching in a program change, you can change the patches after your performance is recorded. You might want to adjust the timbre of each patch with the sound controls on the synthesizer in order to create unusual sounds.

2. Record the First Synthesizer Track

Select your first patch on your synthesizer, and set your synth to omni on-mono mode. This makes it respond monophonically to all channels; it can play only one note at a time—not chords. Set the synth to MIDI channel 1.

Set the desired metronome tempo on your sequencer. Then set your sequencer to record on track 1 and set track 1 to MIDI channel 1. Hit the record key(s) on your computer keyboard.

3. Play the Recording

4. Punch In-Out to Correct Mistakes

5. Record Overdubs on Other Tracks

With your first track recorded and corrected, you're ready to record other tracks. Set your sequencer to record only track 2 and set track 2 to MIDI channel 2. On your synthesizer, select the patch you want to use, and set it to MIDI channel 2.

Then hit the record key(s) on your computer keyboard. Play the new part on the synthesizer while listening to your prerecorded track 1 played by the synthesizer.

Here's an example. Let's say you've just recorded a piano part into the sequencer on track 1. Then you go back to the start of the sequence, play the piano part, and add a bass line on track 2 in sync with the piano. The bass notes are stored separately in memory. Then you go back to the top and add a flute on track 3.

In short, you record a different patch on each track in the sequencer, and play back the recording through the multitimbral synthesizer, which plays all the patches simultaneously. Or, if necessary, use several synths—one for each part. Set each track to a different MIDI channel and each instrument to a corresponding channel.

Suppose you have only one synthesizer, and it can play only four patches at once. There's no point in recording more than four tracks, because you'll only hear four tracks playing. But there's an exception: you might record different performances of the same patch on more than four tracks and then choose the four best tracks.

6. Bounce Tracks

What if your sequencer records eight tracks, but you want to play ten patches at once with several synths? You can make more tracks available by bouncing tracks: merging two tracks into one, thus freeing one of those tracks. Hit the bounce key(s) on your computer keyboard. Type in the source track and destination track (indicate to which track you want to bounce). In a few seconds, the bounce is accomplished.

You can only bounce one track at a time. You can bounce a track into a prerecorded track, however, without erasing the prerecorded track. The prerecorded track and bounced track will merge. Unlike bouncing with a tape deck, there is no generation loss (no loss of sound quality) when you bounce with a sequencer.

You can record program changes on a separate track and later bounce that track to a performance track. The performance track and program-change track will merge into one track.

7. Edit the Composition

8. Mix the Tracks

Now that your song is recorded and arranged, you'll want to adjust the relative volumes of the tracks to achieve a pleasing balance.

If your multitimbral synthesizer doesn't have separate outputs for each patch, you have to adjust the mix at the sequencer. To do this, adjust the volume (key-velocity scaling) of each track by hitting the appropriate computer keys. This only works if your keyboard is velocity sensitive. After you've adjusted the volume of each track in this way, hit the play key(s) on your computer keyboard to play the sequencer. The mix of patches will play on your synth.

If your synth has several individual outputs—one for each patch—connect them to a mixer and set up a stereo mix with panning and effects.

9. Record the Mix

If your synth has a single output (mono or stereo), record the mix off that output, using your 2-track tape recorder. If your synth has several individual outputs connected to a mixer, record off the mixer stereo output.

Using Effects

Effects are an important part of a mix. With a multitrack tape recorder, effects can be added during recording or during mixdown. But with a sequencer, effects can be added only during mixdown because effects are audio signals, which sequencers can't record.

Let's say that while using a multitimbral synthesizer you want to add a different effect to each patch. Whether you can do this depends on your synthesizer. If your synth has a separate output for each patch, you can use a different effect on each patch. But if your synth has only a single output (mono or stereo) and you run it through an effects device, the same effect will be on all the patches.

If your song includes program changes (patch changes), you can give each patch a different effect. Set up a MIDI multi-effects processor so that each synth program change corresponds to the desired effect; when the synth program changes, the effect will change, too.

What if you want the effect, but not the synth patch, to change during a mix? Reserve a track and channel just for effects program changes. You won't hear these program changes in your synth, but you will hear the effects change. During a mixdown, it's usually easier to change effects automatically with your sequencer, rather than manually.

If your synth is a sampling keyboard, each sample could have reverberation or some other effect already on it; in which case, each sample can have a different effect. The effect is not recorded in the sequencer; rather, the effect is part of the sampled sound. Note that the sampled reverberation will cut off every time you play a new note. Although this sounds unnatural, you can use it for special effect.

We said earlier that effects can be recorded on multitrack tape. If an effect is an integral part of the sound of an instrument, it's probably best to record it with the instrument on the multitrack tape. But if the effect is overall ambience or reverb (to put the band in a concert hall), then it's best to add it during mixdown.

Recording a Synthesizer and Drum Machine on Tape

Using a 4-track recorder-mixer, you can record a complete performance without MIDI and without a sequencer. Here's how:

1. Plug your drum machine into track 4 and record the drum part for the song (including the count-off).
2. Plug your synth into track 3. While listening to the drum track, play a synth bass line and record it on track 3.
3. Mix tracks 3 and 4 to track 1. This frees tracks 3 and 4 for more overdubs.
4. Plug your synth into track 2. While listening to track 1 (drums and bass), record a synth Rhodes piano (or whatever) on track 2.
5. Plug your synth into track 3. While listening to tracks 1 and 2 (drum, bass, and Rhodes), record a synthesized piccolo (or whatever) on track 3.

6. Bounce tracks 1, 2, and 3 to track 4. This frees tracks 1, 2, and 3 for more overdubs.
7. Plug your synth into track 3. While listening to track 4 (drums, bass, Rhodes, piccolo), record a synth organ on track 3.
8. Plug your synth into track 2. While listening to tracks 3 and 4 (drums, bass, Rhodes, piccolo, organ), record a synth sweep at the intro of the song.
9. Record hand claps on track 1. Finally, mix down the four tracks to 2-track stereo, and record the mix.

Recording with a Complete MIDI System Plus Tape

Here's a suggested procedure for recording with a complete workstation: a recorder-mixer, a synth with a built-in 8-track sequencer, a drum machine, and effects. Figure 12-7 shows the hookup. Plug the MIDI equipment audio outputs and the tape-track outputs into a mixer, and monitor the mixer's output.

Now that the system is set up, proceed as suggested here:

1. On the tape recorder's track 4, adjust the recording level for the sync tone. Use a level that produces correct synchronization (usually around −4 VU). If you can switch off the noise reduction for the sync track, do so.
2. Set the sequencer to run on its internal clock at the desired tempo.
3. Start recording the FSK leader tone on track 4.

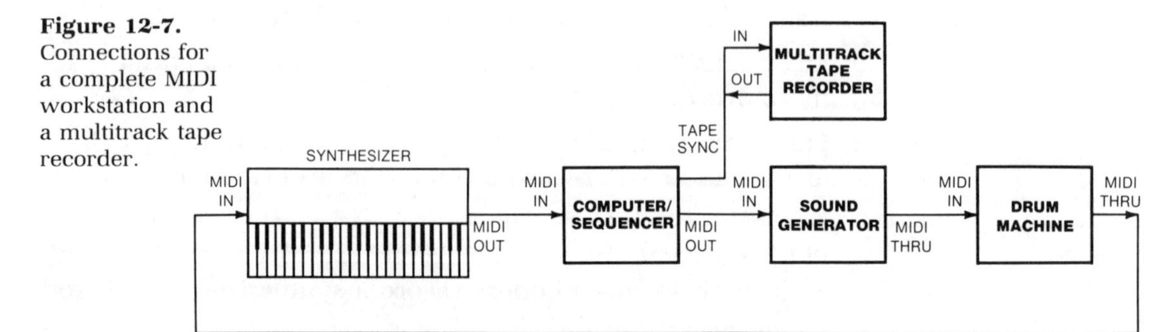

Figure 12-7. Connections for a complete MIDI workstation and a multitrack tape recorder.

4. About 20 seconds later, hit the play key on your sequencer. You'll be recording ("striping") the FSK tape-sync tone on track 4 of the recorder-mixer. This will be your master clock.
5. Let the sequencer run for the duration of the song, plus 30 seconds more for possible additions. While striping the tone, do not record any musical material because it will not sync with material recorded later.
6. Program the drum tracks into the drum machine.
7. Develop your instrumental arrangement using the synthesizer and its sequencer.
8. Set the sequencer and drum machine to external clock mode.
9. Start the tape early in the sync leader tone.
10. After the tape speed has stabilized, press the play key on your sequencer. When the tape-sync tone starts, the sequencer will start playing. It will drive the drum machine and synth at the same tempo.
11. While listening to the synth and drum sequences playing "live" through headphones, record any vocals and non-MIDI instruments on tape. Since you have three tracks available, you could record, say, lead vocal on track 1, harmony vocal on track 2, and sax on track 3.
12. After all your tracks are recorded, set up a mix with panning and effects. Play the song several times to perfect the mix. When you're satisfied with the mix, record it on the 2-track. You'll have three tape tracks, probably eight synth tracks, plus stereo drum tracks, with stereo effects—all first generation! And if you wish, you can modify each track of the synth sequence, changing notes or patches as desired.

Automated Mixdown

When you're doing a mixdown, it can be hard to remember all the fader changes. But there is help. A MIDI-equipped mixer permits computer-controlled (automated) mixdowns via a MIDI signal between the mixer and computer. The computer remembers your fader changes and resets the faders automatically as the song plays. Two such automation systems are the MegaMix system by Musically Intelligent Devices and the MidiMation system by J. L. Cooper Electronics. You can even buy an automated mixer in a single package, such as the MixMate by J. L. Cooper Electronics.

Miscellaneous Tips

To make the music more human sounding, try not quantizing some parts (other than bass and drums), and avoid too many loops on the drum tracks.

Understanding all the technical operations and idiosyncrasies of your equipment can be a formidable task. It helps to read the instruction manuals thoroughly and to simplify them into step-by-step procedures for various operations. If you have questions, call the technical service people at the manufacturer of your equipment—there may be errors or omissions in the instructions.

Working on Arrangements

Musical composition is beyond the scope of this book, but is well covered in many other texts. A recommended tutorial is *Arranging and Recording Electronic Keyboards* by Vinnie Martucci, published by Homespun Tapes. It includes six cassettes and a booklet.

13 Uses for Your Demo Tape

You'll find many ways to use your demo tape. You could send a copy to friends and relatives who want to hear what you're doing. Some musicians record demos as contest entries. You might use your home recordings to document your musical progress and to help you remember your arrangements. Or you can give the tape to band members so that they can learn your songs at home.

There are more commercial possibilities. Play your tape for potential group managers to see whether they are interested in promoting your act. To get jobs, play the demo to club owners who feature music like yours. Your demo tape can get you into studio recording. Maybe you're a great drum programmer or session player or composer of jingles. Making a demo can show your marketable studio skills.

Your home demo is useful even if you've already booked a session at a professional studio. Take your demo into the studio and copy it on a studio recorder containing 8 to 24 tracks; your home-recorded tracks become the foundation of a full-blown studio production. Or you may want to give your record producer a finished 2-track mix to help generate ideas on producing your music. By listening to your tape, a producer can tell what kind of mix you prefer and can suggest special effects, instruments to add or delete, new arrangements, and so on.

You may want to bring your demo tape to booking agencies. A booking agent finds you gigs by playing your demo tape to club owners or individual clients. To find booking agencies, look under "Musicians" or "Entertainment Bureaus" in the Yellow Pages, or ask musician friends for recommendations. The demo tape that you use to get gigs should be a live-performance recording (a work tape) rather than a highly processed studio demo.

A polished home demo might even get you a recording contract. Send your tapes to the A&R departments of several record labels who feature music similar to yours. The names and addresses of record companies can be found in your public library's reference section in the book *Songwriter's Market.* Corporate directories include the larger record companies and you can get addresses from record jackets. Be sure to duplicate a sufficient number of cassette copies, either by yourself or by a cassette-duplication company (found in the Yellow Pages under "Recording Service.")

Whether you're sending tape to agents or record companies, send one cassette only, containing just enough tunes to show the range of material you can do—a maximum of five songs. Executives have little time to audition tapes. In fact, it's a good idea to begin your tape with short segments of your best work, say, 20 seconds of the best part of each song.

Most record-company A & R people do listen to submitted cassettes, and return a note letting you know whether they're interested in having you do a recording session. They will pay for the session and for promotion of the recording if it's released commercially. Unfortunately, the chances of this happening with a Top-40 label are slim unless your material is competent *and* commercial and has a hook.

Don't send your demo tape to disk jockeys or program directors at radio stations. They expect you to have album-length recordings already in record stores and a performance schedule.

Protecting Your Rights

Before sending the demo, obtain a copyright to protect your original songs from being stolen. Write to the Register of Copyrights and request Form PA, Application for Copyright Registration for a Work of the Performing Arts. Request a Form PA for each song you want to register. Also ask for a copy of the Copyright Law and copies of any regulations or publications about copyright of musical works. Take the time to read these so you understand your rights and responsibilities.

For each song you want to copyright, send the completed Form PA and $10 to the Register of Copyrights, Library of Congress, Washington, DC 20559. Include a demo tape containing all the songs you wish to copyright. Although the tape is sufficient, some lawyers recommend sending a lead sheet (the song's lyrics, chords, and melody in musical notation) for each song. A lead sheet can be written for you by the head of the music department in a college or high school or by a member of the local symphony orchestra. Your music union local might be able to suggest a professional lead-sheet copyist. If your song is recorded with a sequencer, a notation program can print the sequence in standard musical notation.

What to Send to a Producer

On the cassette label, type or print neatly your name (if you're a solo), the name of your band, the date, a copyright notice, and, if you have room, the song titles. Also note the type of noise reduction used (Dolby B is recommended because most cassette decks have it).

Along with the cassette, send a "promo" pack including a professional photo (8″ × 10″ black and white glossy), posters or flyers, and a typed letter describing the following:

- For booking agents, mention your intended audience(s): rock club, jazz club, concerts, parties, weddings, and so on.
- Your music's category(ies): rock, pop, country, classical, blues, jazz, fusion, heavy metal, old-time, country rock, ethnic, R&B, new wave, new age, experimental, and so on.
- What makes your music unique or special—its mood, social or political commentary, technical features, and so on.
- A list of places you've played, if any.
- Copies of magazine and newspaper reviews of your records or performances, if any.
- Name, address, and phone number of the person to contact.

Doing an Album

You may want to record a whole album's worth of material, and promote and sell the record yourself. You could sell cassettes at concerts or gigs or sell them to record stores. An excellent book on this type of independent recording and promotion is *How to Make and Sell Your Own Record* by Diane Sward Rapaport, Headlands Press, available from Music Sales Corp., 799 Broadway, New York, NY 10003.

If you'd like to engineer an album yourself, a good how-to book is *Introduction to Professional Recording Techniques* by the author, published by Howard W. Sams & Company. (See page ii of this book for ordering information). It is also available from The Mix Bookshelf Catalog, 6400 Hollis St., #12, Emeryville, CA 94608. The book goes into more technical detail than this one and covers these additional topics:

- more on studio acoustics
- monitoring
- more on hum prevention
- microphones
- microphone-technique basics
- more on tape recording
- creative sonic effects
- recording the spoken word
- on-location recording of classical music
- judging sound quality of classical-music recordings
- the decibel
- SMPTE time code

This book has focused on efficient ways to make an effective demo tape at home. The better your recording is, the better your chances are of getting jobs or recording contracts. With a professional-sounding production, you can confidently send your tape to others who may enhance your musical career.

A Training Your Hearing

Here's a series of experiments that can train your hearing. You'll learn not only what to listen for in recordings, but also how to achieve various sounds through recording techniques. Announce on tape each experiment, and take notes on the results.

We'll start with very simple methods and work up to complex ones, a route like that taken by many professional engineers in the course of their careers.

Recording a Soloist with One Microphone

Place a microphone 1′ from a musical instrument or vocalist and plug the mic into a tape recorder (as in Figure A-1). Plug in headphones to monitor the recording and playback.

Although this setup is simple, there's a lot to be learned from various experiments:

Figure A-1. Setup for one-mic recording experiments.

1. Make recordings at −20 VU, 0 VU, and +6 VU (far into the red). You'll hear how excessive levels cause distortion and too-low levels make tape hiss audible. Note that percussive instruments require lower recording levels for undistorted recordings than do nonpercussive instruments.
2. Keep the recording level at 0 VU, and make recordings at 6″, 1′, 2′, 4′, and 10′. You'll learn how miking distance affects the amount of room reverberation heard in the recording. Distant placement sounds distant; close placement sounds close.
3. At a miking distance of 1′, make recordings with several different microphones (one at a time) at the same recording level. Listen to the tonal differences of the various microphones. Make a note of what you heard for future reference. Repeat this experiment with various musical instruments.
4. For this experiment, you need a single-D cardioid mic and a multiple-D cardioid mic. "D" refers to the distance between the diaphragm and the rear sound entry. A single-D microphone has a single sound entry (screened window) behind the diaphragm; a multiple-D microphone has more than one sound entry at differing distances behind the diaphragm.

   An example of a single-D cardioid is a Shure SM-57 or SM-58. Hold such a microphone 2′ from your mouth. While maintaining a constant recording level, talk into the mic and move it slowly toward your mouth until it's touching your lips. You'll hear the bass increase as the mic gets closer—a phenomenon called *proximity effect.*

   Repeat this experiment with a multiple-D cardioid mic (such as an Electro-Voice RE-18) or an omnidirectional mic. The tonal balance will stay relatively constant with miking distance.
5. Record a loud instrument (such as a kick drum, guitar amp, or screaming voice) 1″ from the mic at a −8-VU recording level. Play back the tape. Even though you recorded at a moderate level, you're likely to hear distortion caused by the microphone signal overloading the mic preamp in the tape recorder. Insert a pad in the microphone cable (such as a plug-in pad with 20-dB loss, as was shown in Figure 10-2) and repeat the recording. The distortion should be eliminated.

6. Record the output of a cardioid microphone. Talk into the mic from all sides while maintaining a constant distance from it. Play back the tape. Your reproduced voice will be loudest in front of the mic (on-axis) and softest behind it. Note that off-axis rejection of sound is best in an open, nonreflective area. Off-axis rejection is poor when the front of the mic is aiming at a nearby wall that reflects sound into the front.
7. Repeat Experiment 6 with an omnidirectional microphone. The overall level will stay constant when you walk around the mic, but the high frequencies ("s" sounds) may become duller off-axis. You'll find that miniature microphones have very little change in sound off-axis.
8. Using one microphone, place the mic at various positions around the sound source, keeping a constant distance (say 1′), and record the result. You'll hear how microphone placement affects the tonal balance. Repeat this for various instruments and note the results. This is one way to become more proficient in microphone techniques.
9. Record an electric guitar with a direct box connected to (1) the guitar, (2) the output of the effects boxes, and (3) the amplifier speaker jack. Compare these recordings to one made with a microphone near the amplifier speaker.

Recognizing Phase Cancellations

Place the microphone near a hard reflective surface, such as a table top, and make a recording. Start with the mic 1′ above the table top, and, repeating the phrase "Sally's sister," slowly lower the microphone to the surface. During playback you'll hear a filtered, swishy tonal change called *comb filtering,* which is caused by phase interference between the direct and reflected waves arriving at the microphone.

For comparison, record with a boundary mic (such as a Crown PZM) placed on the table. Phase cancellations are eliminated.

Recognizing Leakage

Let's move on to another experiment involving two microphones and a mono mixer. Plug the mics into the mixer, and connect the mixer output to your recorder line input (as shown in Figure A-2). Place one microphone close to a snare drum, and place the other mic several feet away. The distant mic represents a microphone intended for another instrument, such as an acoustic guitar.

Set the mixer master volume control ¾ up. Turn up the volume control of the drum microphone and make a recording at −8 VU. You'll hear a clean, tight drum sound.

Now, while recording the drum, gradually turn up the distant mic and mix it with the drum mic. You'll hear the clean drum sound become distant and muddy as you bring up the distant mic. That's caused by leakage of the drum sound into the distant microphone.

Next, turn up the distant mic just enough to make the drum sound muddy. To reduce leakage into the distant microphone, (1) aim the "dead" rear of the distant mic at the drum, (2) place the distant mic in a padded, acoustically dead area, (3) place a large, heavy wooden panel between the drum and distant mic, and (4) place the distant mic at various distances from the drum. Record these experiments and note how the clarity of the drum sound changes each time.

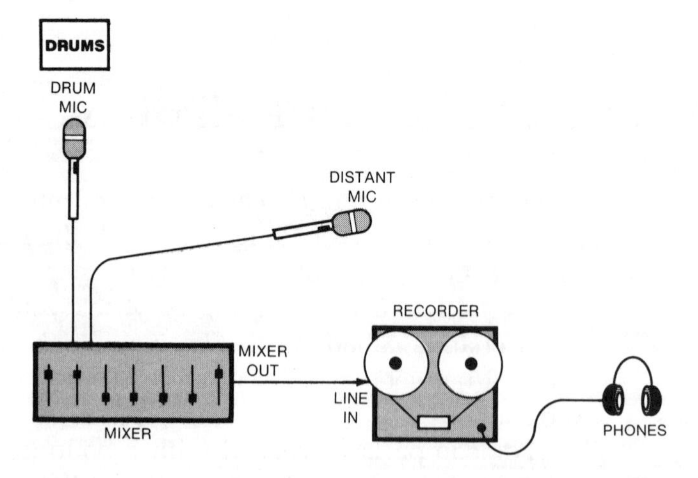

Figure A-2. Setup for leakage experiments.

Hearing Stereo Effects

For this experiment, plug two identical cardioid microphones into a recorder and place them as shown in Figure A-3. Record yourself speaking while you walk around the microphone pair at a constant distance. Say "I'm speaking in position A; I'm speaking in position B," and so on. Play back this stereo recording over two loudspeakers. Sit equidistant from the speakers, as far from them as they are apart, as shown in Figure A-4. You'll hear yourself speaking from various positions between the speakers.

In making the recording, try these variations:

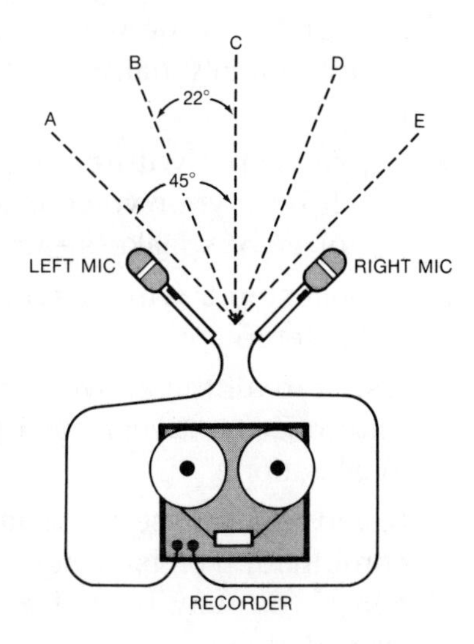

Figure A-3. Recording setup for stereo localization experiment.

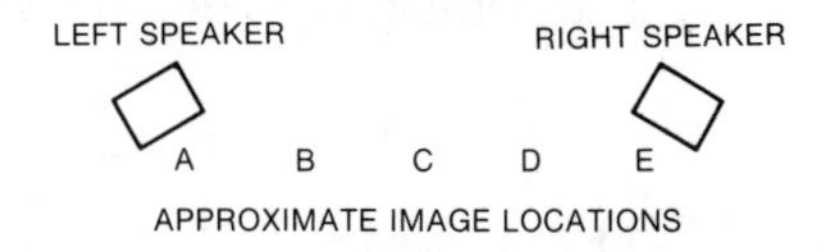

Figure A-4. Playback setup for stereo localization experiment.

1. Angle the cardioid microphones 110° apart (55° to the left and right of center) and spaced 7″ apart horizontally. You'll hear fairly accurate stereo localization and sharp imaging.
2. Try various anglings and spacings of the two cardioid mics, and record and play back the results. Try eliminating the spacing—put the grille of one mic directly over the grille of the other, and angle the mics apart. Note the results.
3. Place two omnidirectional mics 3′ apart, aiming straight ahead, and record yourself as described above. In positions A or E (Figure A-4), your voice should be reproduced from the left or right speaker. In position C, your voice should be reproduced midway between the speakers (that is, straight ahead). In positions B or D, your voice should be approximately halfway off- center, but without pinpoint imaging.
4. Repeat step 3 with two omni mics spaced 10′ apart. Now you'll hear your voice in positions B and D coming from the left or right speakers—an exaggerated separation effect.
5. Repeat step 3 with two omni mics 6″ apart. You'll hear very little stereo spread.
6. Listen to all the above recordings on headphones. Note the differences between headphone localization and speaker localization.
7. Listen to all these recordings in mono (both channels combined). For techniques in which the mics were spaced apart, you may hear the tone quality change as the recorded voice moves around the microphone pair. This is due to phase cancellations between the mics.

Doing a Live Mono Mix

Refer to Figure A-5 for this next experiment. Plug several microphones into a mixer without equalization or effects. Use a mono mixer or, if your mixer is stereo, assign all the mics to one channel. Connect the mixer output to a recorder line input.

Place each microphone close to each instrument or voice in a position where your experiments have yielded a natural sound.

Set the mixer master volume ¾ up. Turn up the microphone volume controls. As the instruments are playing, adjust the pad or gain-

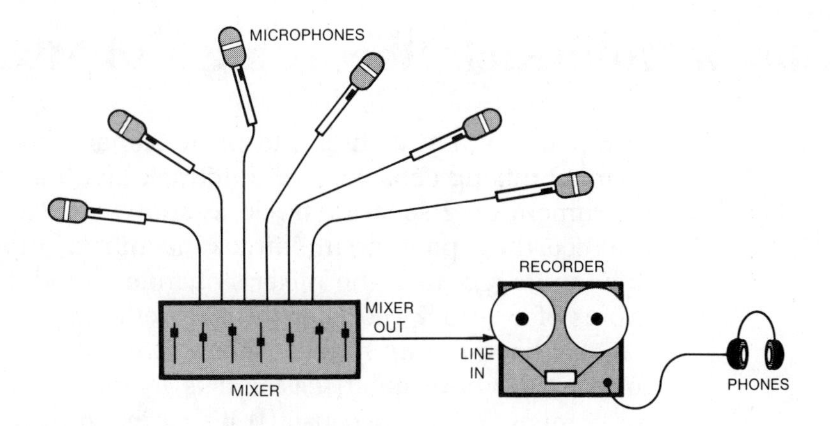

Figure A-5. Setup for a live mono mix.

trim control for each microphone input to prevent input-overload distortion. Many mixers have LEDs that flash when input levels are excessive; turn down the gain trim (increase the attenuation) until the LEDs just go out.

Do a live mono mix of the instruments and vocals, with the mixer meter and recorder meter both peaking around 0 VU. Is anything too loud or too quiet? Can you hear everything? Can you understand the lyrics? Adjust the volume controls for a pleasing balance and record the mix. Play it back and evaluate it.

Try turning up the mixer level until the meter starts banging against the stop, and reduce the tape deck's recording level to 0 VU. Listen for the distortion caused by overloading the recorder input.

Doing a Live Stereo Mix

Let's complicate things a little. Use the same setup as with the mono mix, but use a stereo mixer. Assign instruments to channel 1 (left) or channel 2 (right) or both. Use pan pots to position the instruments in various locations. Normally the kick drum, bass, and lead vocal go to center, while keyboards and rhythm guitar are either in stereo or are split equally left and right.

Note the "unbalanced" effect of placing bass or drums in only one channel. Also note how it's easier to hear each performance of two similar instruments if they are separated spatially.

Monitor the recording alternately in mono and stereo to see if the mix changes. Listen for *center-channel buildup:* in mono, centered instruments and vocals sound louder relative to noncentered instruments than they do in stereo.

Doing a Multitrack Recording and Mixdown

Now we're really getting into professional techniques. Use a multi-channel mixing console and multitrack tape machine to record each instrument on a separate track, as shown in Figure A-6. After doing the recording, plug the multitrack machine's outputs into the mixer's line inputs. Connect the mixer's channel 1 and 2 outputs to the line inputs of a 2-track recorder (or a cassette deck).

Play back the multitrack tape and mix the instruments with the mixer. Note how much easier it is to mix a tape than to mix live instruments. That's because (1) it's easier to hear what you're doing with no live band playing, and (2) you can play the multitrack tape over and over to practice the mix.

Adding Equalization

Using the same setup as before, add EQ to each instrument. Apply extreme boost or cut at various frequencies to hear the effect. Note that the same EQ affects different instruments differently. For example, a boost at 12 kHz affects cymbals but not bass guitar.

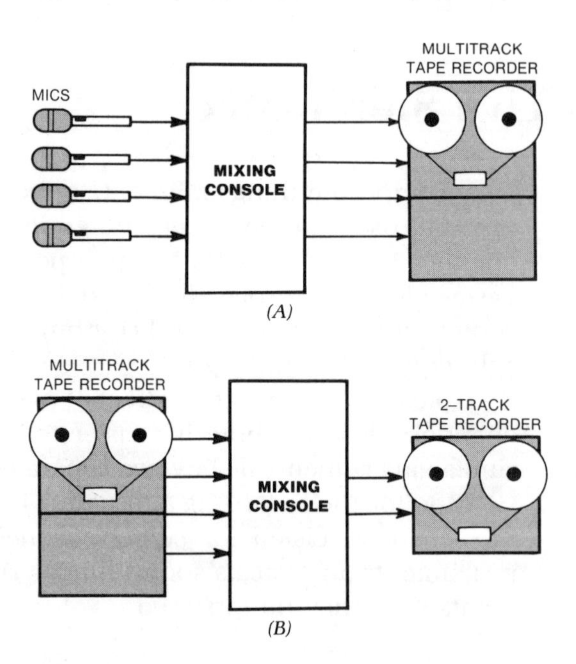

Figure A-6 (A) Multitrack recording and (B) mixdown to 2-track

Apply descriptive terms to the tonal changes, such as "bassy, warm, dull, crisp, present," or "edgy." The next time someone asks you for a particular tonal balance on a certain instrument, you'll have an idea how to get it.

Listen to each instrument alone and set the EQ for the most natural tonal balance. Some instruments may need no EQ at all, depending on the microphone chosen and its placement. Then mix all the instruments together and note whether the EQ needs changing. Often, instruments mask or cover up each other's high frequencies, so extra EQ is needed when instruments are combined in a mix.

Compare recordings made with and without EQ. Ideally, the EQd recordings should sound much better tonally, unless your microphone techniques resulted in a great sound without EQ.

Adding Reverberation

Using the previous setup, connect an external reverb unit between the aux-send and -receive jacks on your mixer. Next, set the aux-return level (if any) about halfway up. Now adjust the aux-send level on each instrument's channel for the desired amount of reverberation.

Usually, the bass and kick drum get little or no reverberation so that they retain their clarity. Note the effect of adding reverb to these instruments. Try adding too much reverb to vocals and note the loss of clarity.

Compare mixes made with and without reverberation. Those with reverberation sound more spacious, as if they were recorded in a concert hall.

If possible, compare a mono to a stereo reverb unit. Pan the mono reverb-return signal to center; pan the stereo reverb-return signals to left and right. Note the increase in spatial realism that stereo adds.

Adding Compression

Since vocals have a wider dynamic range than the instrumental backup, vocals occasionally sound too loud or too quiet. Compressing the vocals keeps their level more constant relative to the instruments.

Plug a compressor between the vocal tape-track output and a console line input. Bring up the vocal volume control, play the tape, and listen to the compressed vocal track alone.

A typical threshold setting on the compressor is −10 dB; a typical compression ratio is 3:1. Start with these settings, and then change them and listen to the effect. Too much compression sounds "squashed" or "forced."

Now mix in the rest of the instruments. The compression should be less audible, and the volume of the vocals should be better controlled. Listen to the effect of compression on kick drum, drums, bass, and lead guitar. Try to verbalize these effects so that you can talk with others about how compression sounds.

Adding Delay

Here's another signal processor to play with. Connect a digital delay to the aux-send and -receive jacks on your mixer. Set the mix control on the delay unit to delay. Play a drum track through the mixer, and turn up its aux send until you hear both the direct and delayed signals in equal proportions. Adjust the delay from 0 seconds to 1 second, and note the effect of various delay settings.

Up to about 20 ms, you'll hear flanging or comb filtering. At about 50 to 100 ms, you'll hear a quick slap echo. Slow repetitions occur around .5 to 1 second.

Turn up the recirculation control (if any) until the echoes repeat. Add a repeating echo to vocals, and note the amount of recirculation needed for a tasteful result.

Summary

To train your hearing, it helps to spend some time with each piece of recording equipment. Get to know the audible effects of different control settings and microphone placements. Compare the sound of various pieces of equipment. Be sure you have recorded the results and announced each experiment on tape for future reference.

By trying these experiments, you can train yourself to hear characteristics in recordings that you may not have noticed before. You'll learn how to achieve all sorts of effects in your own recordings and when you listen closely to commercial recordings, you'll start to perceive microphone techniques, mix balances, EQ, reverb, delay, and so on.

B Basics of Sound

Before you can understand all the devices that pick up and process sound, you need a good grasp of the fundamentals of sound itself. Although you can make a recording without studying the material in this appendix, you'll find it easier to choose and operate recording equipment once you understand some characteristics of sound and signals.

Sound-Wave Creation

To produce sound, most musical instruments vibrate against air molecules. The air molecules pick up the vibration and pass it along as sound waves. When these vibrations strike our ears, we hear sound.

Let's examine how sound waves are created. Suppose a speaker cone in a guitar amp is vibrating—moving rapidly in and out. When the cone moves out, it pushes the adjacent air molecules closer together. This forms a *compression.* When the cone moves in, it pulls the molecules farther apart, forming a *rarefaction.* As illustrated in Figure B-1, the compressions have a higher pressure than normal atmospheric pressure; the rarefactions have a lower pressure than normal.

These disturbances are passed from one molecule to the next in a springlike motion; each molecule vibrates back and forth to pass the wave along. The sound waves travel outward from the sound source at 1,130 feet per second.

At a receiving point, such as an ear or a microphone, the air pressure varies up and down as the disturbance passes by. Figure B-2 graphs how sound pressure varies with time. It fluctuates up and down like a wave; hence the term "sound wave." The high point of

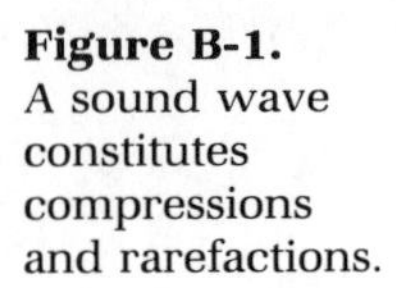

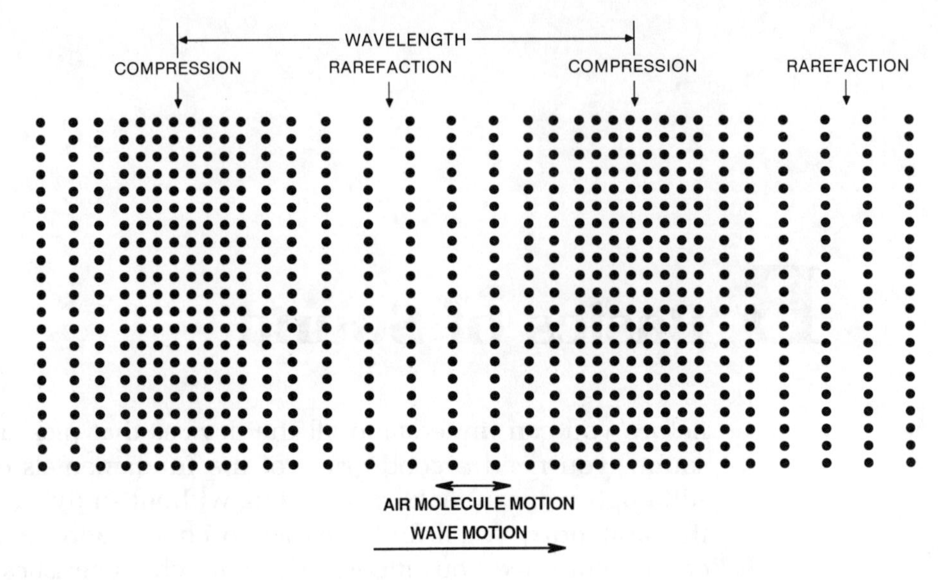

Figure B-1. A sound wave constitutes compressions and rarefactions.

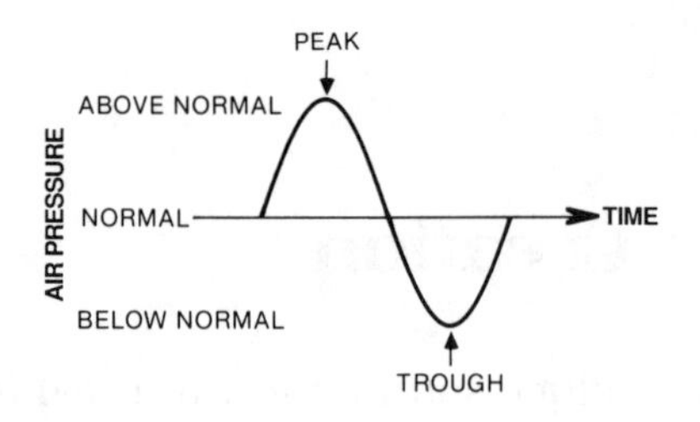

Figure B-2. Sound pressure vs. time in one cycle of a sound wave.

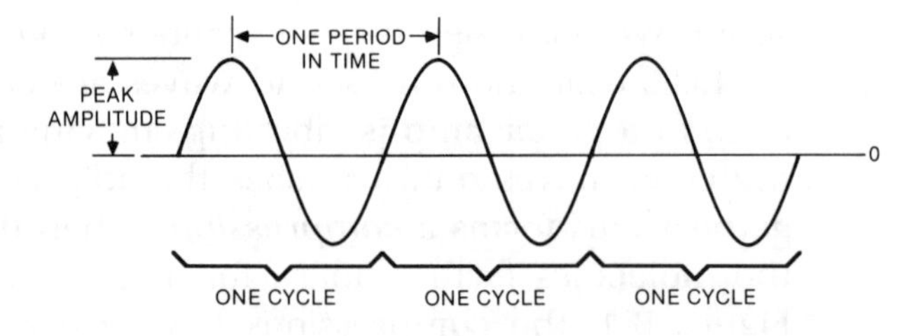

Figure B-3. Three cycles of a wave.

the graph is called a *peak;* the low point is called a *trough.* The horizontal center line of the graph is normal atmospheric pressure.

Characteristics of Sound Waves

Figure B-3 shows three waves in succession. One complete vibration from high to low pressure and back to the starting point is called a *cycle.* The spacing *in time* between the peak of one wave and the peak of the next is called the *period* of the wave. One cycle is one period long.

Amplitude

At any point on the wave, the vertical distance of the wave from the center line is called its *amplitude.* The amplitude of the peak is called the *peak amplitude.* The more intense the vibration, the greater the pressure variations and the greater the peak amplitude. The greater the amplitude, the louder the sound.

Frequency

The sound source (in this case, the loudspeaker) vibrates back and forth many times a second. The number of cycles completed in one second is called *frequency.* The faster the speaker vibrates, the higher the frequency of the sound. Frequency is measured in hertz (Hz), which stands for cycles per second. One-thousand hertz is called 1 kilohertz (kHz). The higher the frequency, the higher the perceived pitch of the sound. Low-frequency tones (say, 100 Hz) are low pitched; high-frequency tones (say, 10,000 Hz, or 10 kHz) are high-pitched. Doubling the frequency raises the pitch one *octave.* Each musical instrument produces a range of frequencies; say, 40 Hz to 5 kHz, or 500 Hz to 15 kHz. Young children can hear frequencies from 20 Hz to 20 kHz, and most adults with good hearing can hear up to 15 kHz or higher.

Wavelength

As a sound wave travels through the air, the physical distance from one peak (compression) to the next is called a *wavelength* (refer to Figure B-1). Low frequencies have long wavelengths (several feet); high frequencies have short wavelengths (a few inches or less).

Phase and Phase Shift

The *phase* of any point on the wave is its degree of progression in the cycle—the beginning, the peak, the trough, or anywhere in between. Phase is measured in degrees, with 360° being one complete cycle. The beginning of a wave is 0°; the peak is 90° (¼ cycle), and the end is 360°. Figure B-4 shows the phase of various points on the wave.

If there are two identical waves, but one is delayed with respect to the other, a *phase shift* occurs between the two waves. The more delay, the more phase shift. Phase shift is also measured in degrees. Figure B-5 shows two waves separated by 90° (¼ cycle) of phase shift.

When there is a 180° phase shift between two identical waves, the peak of one wave coincides with the trough of another. If these two waves are combined, they cancel out, a phenomenon called *phase cancellation.* We often hear it when two microphones picking up the same source are mixed together. We also hear it when a microphone is near a hard reflective surface, so that the microphone picks up both direct and reflected sounds.

Harmonic Content

The type of wave shown in Figure B-2 is a *sine wave.* It is a pure tone of a single frequency, such as is produced by a tone generator. However, most musical tones have a complex waveform, with more than one frequency component. Yet no matter how complex, all sounds are combinations of sine waves of different frequencies and amplitudes.

Figure B-6 shows sine waves of three frequencies combined to form a complex wave. The amplitudes of the various waves are added

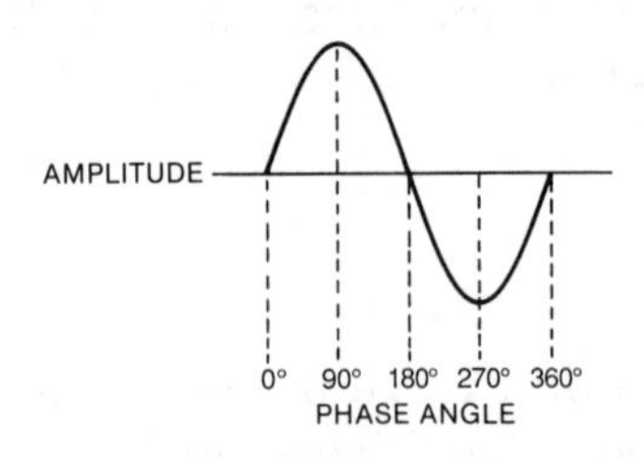

Figure B-4. The phase of various points on a wave.

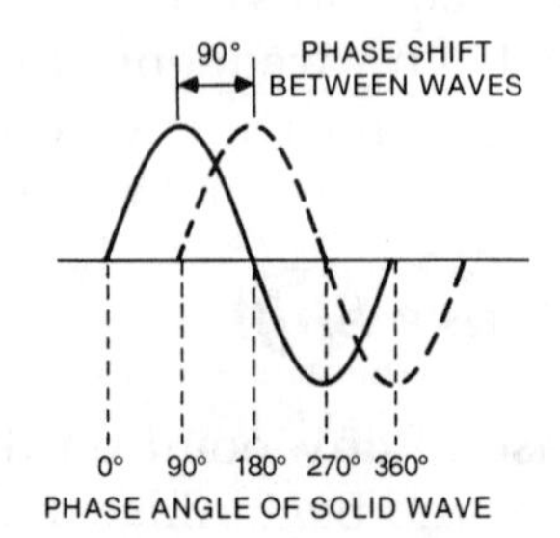

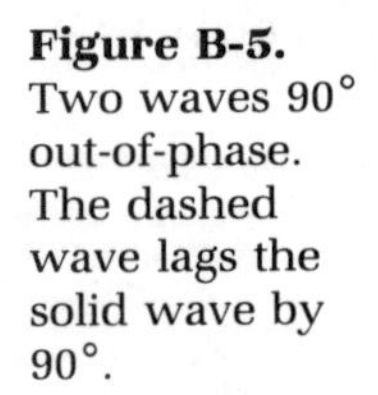

Figure B-5. Two waves 90° out-of-phase. The dashed wave lags the solid wave by 90°.

Figure B-6. Addition of fundamental and two harmonics to form a complex waveform.

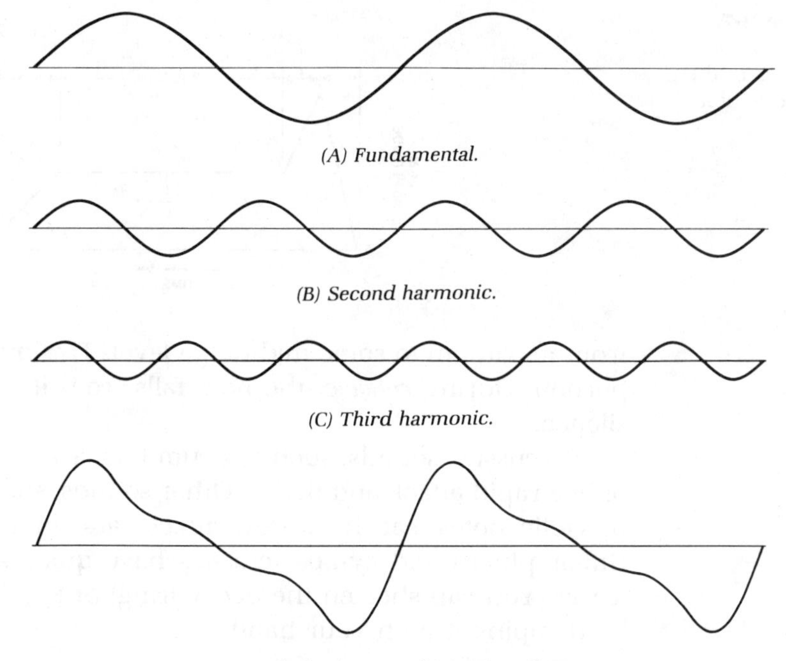

(A) Fundamental.

(B) Second harmonic.

(C) Third harmonic.

(D) Algebraic sum of (A), (B), and (C).

algebraically at the same point in time to obtain the final complex waveform. The lowest frequency in a complex wave is called the *fundamental frequency.* It determines the pitch of the sound. Higher frequencies in a complex wave are called *overtones* or *upper partials.* Overtones that are integral multiples of the fundamental frequency are called *harmonics.* For example, if the fundamental frequency is 200 Hz, the second harmonic is 400 Hz; the third harmonic is 600 Hz, and so on.

The number of harmonics and their amplitudes relative to the fundamental partly determine the tone quality or *timbre* of a sound. They identify the sound as being from a trumpet, piano, organ, voice, and so on. *Noise* (such as tape hiss) contains all frequencies and has an irregular, nonperiodic waveform.

Envelope

Another identifying characteristic of a sound is its *envelope:* the rise and fall in volume of one note. The envelope connects the peaks of successive waves that make up a note. An envelope has four sections, as shown in Figure B-7: attack, decay, sustain, and release. During the *attack,* a note rises from silence to its maximum volume. Then it *decays*

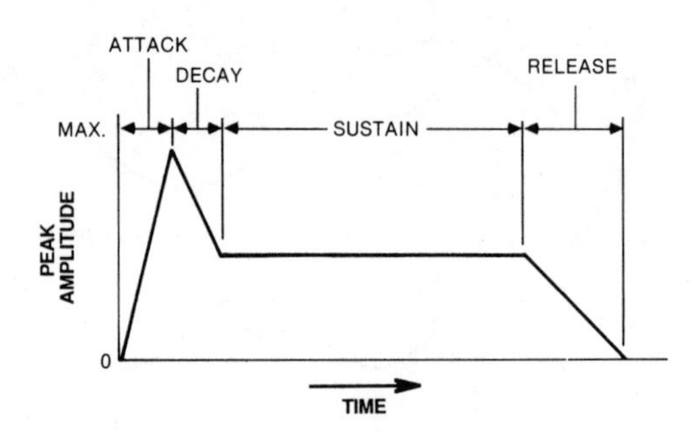

Figure B-7. The four portions of the envelope of a note.

from maximum to some mid-range level. This mid-level is the *sustain* portion. During *release,* the note falls from its sustain level back to silence.

Percussive sounds, such as drum hits, are so short that they have only a rapid attack and decay. Other sounds, such as sustained organ or violin notes, can have slow attacks and high-level, long sustains. Guitar plucks and cymbal crashes have quick attacks and slow releases. You can shorten the decay (ring) of a guitar string or cymbal by damping it with your hand.

Behavior of Sound in Rooms

So far we've covered the characteristics of sound waves traveling in open space. But since most music is recorded in rooms, we need to understand how room surfaces affect sound.

Echoes

As sound travels outward in all directions, some travels directly to the listener (or to a microphone) and is called *direct sound.* The rest strikes the walls, ceiling, floor, and furnishings of the recording room. At those surfaces, some of the sound energy is absorbed, some is transmitted through the surface, and the rest is reflected back into the room.

Since sound waves take time to travel (about 1 ft/ms), the reflected sound arrives after the direct sound reaches the listener. The delayed arrival of a reflected sound causes a repetition (*echo*) of the original sound (see Figure B-8). In large rooms we sometimes hear discrete single echoes; in small rooms we often hear a short, rapid succession of echoes called *flutter echoes.*

Figure B-8. Echoes.

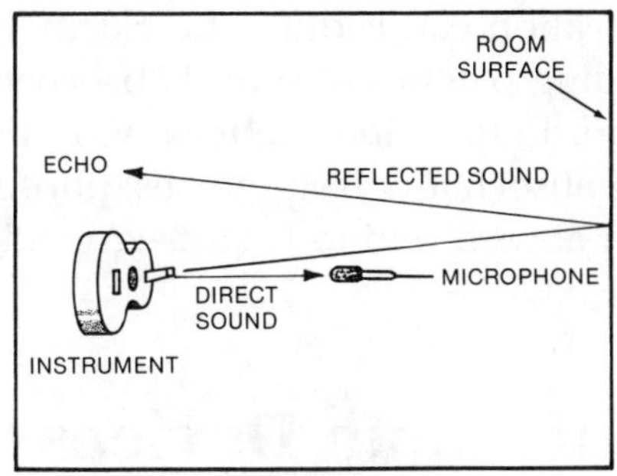

(A) *Echo formation.*

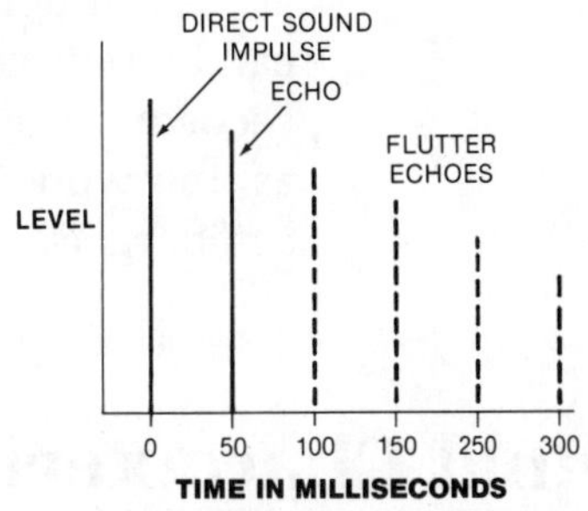

(B) *Intensity vs. time of direct sound and its echoes.*

Figure B-9. Reverberation.

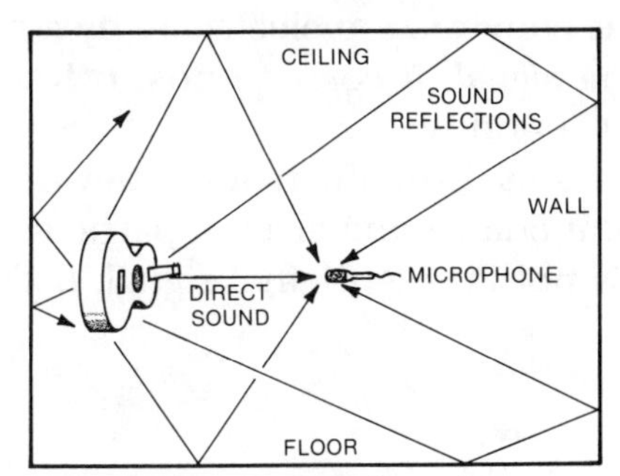

(A) *Reverberation formation.*

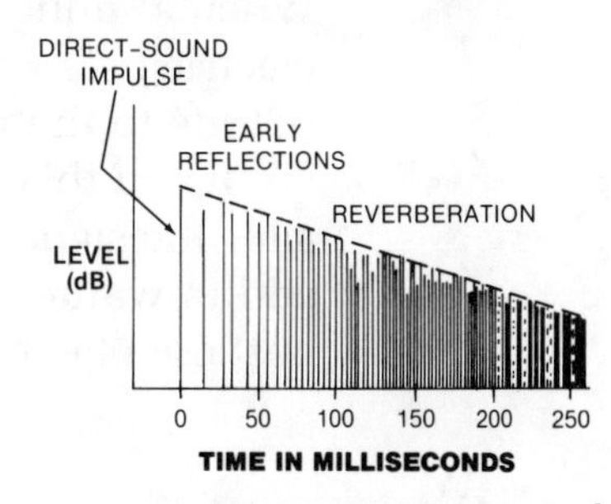

(B) *Intensity vs. time of direct sound, early reflections, and reverberation.*

Echoes can reduce the clarity of a recording, but they can be reduced or eliminated by adding patches of sound-absorbing material on the walls, ceiling, and floor. Some of these materials are fiberglass insulation, thick blankets, curtains, or carpeting.

Reverberation

Sound reflects not just once but many times from room surfaces. These reflections sustain the sound in the room for a short time even after the original sound is no longer audible. This phenomenon is called *reverberation*—the persistence of sound in a room after the original sound has ceased. For example, reverberation is the sound you hear just after you shout in an empty gymnasium. The sound of your shout persists in the room and gradually dies away (decays).

In physical terms, reverberation is a series of echoes which decreases in intensity over time and is eventually absorbed by the inner surfaces of a room. The echoes are so closely spaced in time as to seem to merge into a single continuous sound. The timing of the echoes is random, and they increase in number as they decay. Figure B-9 shows reverberation as a decay of room reflections.

Too much reverberation can reduce the clarity of a recording. It can be reduced by adding patches of sound-absorbent material to the recording room, as you do to reduce echoes. Vibrating surfaces such as thin wood paneling absorb low frequencies; porous materials such as fiberglass insulation absorb mid-to-high frequencies.

Signal Characteristics of Audio Devices

When a sound wave is converted to electricity by a microphone, this electricity is called the *signal.* It corresponds in frequency and amplitude to the original sound wave.

When this signal passes through an audio device, the device may alter the signal. It might change the level of particular frequencies or add unwanted sounds not in the original signal. Let's examine some of these effects.

Frequency Response

The frequency response of an audio device is the range of frequencies it reproduces at an equal level (within a tolerance, such as ±3 dB). Frequency response is often plotted on a graph showing the signal level versus frequency (as shown in Figure B-10). Signal level is measured in decibels, while frequency in measured in hertz. In the illustration the frequency response is 50 Hz to 12 kHz ±3 dB. This means the audio device with this response passes all frequencies from 50 Hz to 12 kHz at a nearly equal level—within 3 dB. Low-pitched sounds and high-pitched sounds are reproduced equally well (within 3 dB of

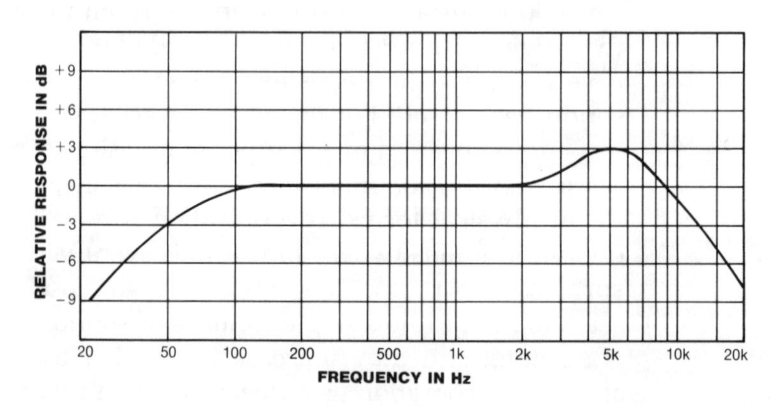

Figure B-10. Example of a frequency response.

each other). The response is down 3 dB at 50 Hz and 12 kHz and is up 3 dB at 5 kHz.

If the response is the same at all frequencies within the specified range, it forms a horizontal straight line and is called a *flat* frequency response (as is shown in Figure B-11).

For high-fidelity reproduction, an audio device should have a flat frequency response over the range of frequencies that the musical instrument or voice produces. That way, fundamentals and harmonics are reproduced in the same proportion as they occurred in the original sound source. Since the ratio of fundamentals and harmonics affects the tone quality or timbre, preserving the original ratio also preserves the original tone quality.

If the frequency response decreases (rolls off) at high frequencies (as on the right side of Figure B-10), then the upper harmonics are weakened and the result is a dull sound. If the response rolls off at low frequencies (as on the left side of Figure B-10), then the fundamentals are weakened and the result is a thin sound. Note that if either rolloff occurs beyond the range of frequencies that an instrument produces, then the rolloff is inaudible.

In general, the wider the frequency response (the greater the range of frequencies that are reproduced equally), the higher the fidelity. A wide frequency response generally results in accurate reproduction. A frequency response of 200–8,000 Hz (±3 dB) is narrow (poor fidelity), a response of 80 Hz to 12 kHz is wider (better fidelity), and a response of 20 Hz to 20 kHz is widest (best fidelity). Also, the flatter the frequency response, the greater the fidelity or accuracy. A response deviation of ±3 dB is good, ±2 dB is better, and ±1 dB is excellent.

Note that the frequency response of an audio device may be intentionally altered from flat for special effect. Also, a microphone may

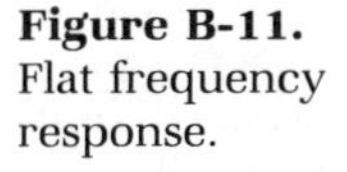
Figure B-11. Flat frequency response.

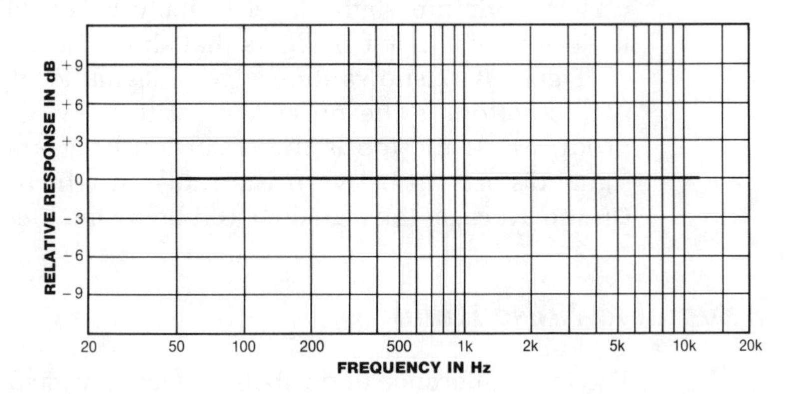

sometimes sound best with a nonflat response. In general, though, a wide, flat response results in high-fidelity reproduction.

Noise

Every audio component produces noise—a rushing sound like wind in trees. Noise in a recording is undesirable. It can be made less audible by keeping the signal level relatively high. If the signal in an audio device is very low, you have to turn up the listening volume in order to hear the signal well. But turning up the volume of the signal also turns up the volume of the noise, so that you hear noise along with the signal. If the signal level is high to start with, you don't have to turn up the listening level so high, so the noise remains in the background.

Distortion

If the signal level is too high, distortion occurs and gives a gritty, grainy quality to the signal. This type of distortion is sometimes called *clipping* because the extremes of the waveform are clipped off. To hear distortion, simply record a signal at a very high recording level (with the meters going well into the red area) and play it back.

Optimum Signal Level

We want the signal level high enough to cover the noise, but low enough to avoid distortion. Every audio component works best at a certain optimum signal level (usually called 0), and this is usually indicated by a meter or lights that show the signal level.

Figure B-12 shows the range of signal levels in an audio device. At the bottom is the noise floor of the device—the level of noise it produces. At the top is the distortion level—the point at which the signal distorts. In between is a range in which the signal may vary. On the average, the signal should be maintained around the 0 point.

Signal-to-Noise Ratio

The level difference in decibels between the signal level and the noise floor is called the signal-to-noise (S/N) ratio (shown in Figure B-12).

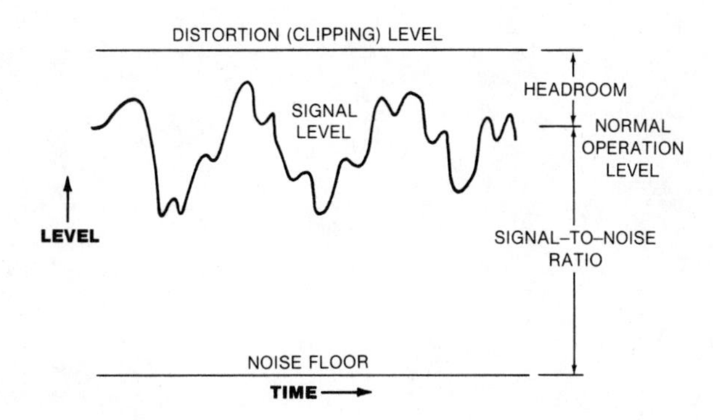

Figure B-12. The range of signal levels in an audio device.

The higher the S/N ratio, the cleaner the sound. An S/N ratio of 50 dB is fair, 60 dB is good, and 70 dB or greater is excellent.

To illustrate S/N ratio, imagine a person yelling a message over the sound of a train. The message being yelled is the signal and the noise is the train. The louder the message, or the quieter the train, the greater the S/N ratio. And the greater the S/N ratio, the clearer is the message (the signal).

Headroom

The level difference in decibels between the nominal (normal) signal level and the distortion level is called *headroom* (shown in Figure B-12). The greater the headroom, the greater the signal level the device can pass without running into distortion. If an audio device has a lot of headroom, it can pass peaks in the waveform without clipping them.

Conclusion

We've covered the basics of sound waves, sound behavior in rooms, and signal characteristics. With this foundation, you're better prepared to choose and operate equipment for your home recording system.

C Reference Sources

These books and magazines were valuable resources for this author and are recommended to anyone seeking further education in recording technology.

Books and Magazines

The following books are available from The Mix Bookshelf Catalog, The Recording Industry Resource Center, 6400 Hollis St., #12, Emeryville, CA 94608.

Anderton, Craig. *Home Recording for Musicians.* New York: GPI Publications, Music Sales Corp., 1978.

_____*Midi for Musicians.* New York: GPI Publications, Music Sales Corp., 1986.

Bartlett, Bruce. *Introduction to Professional Recording Techniques.* Indianapolis, Indiana: Howard W. Sams & Co., 1987.

Borwick, John. *Sound Recording Practice,* 2d Ed. London: Oxford University Press, 1980.

Clifford, Martin. *Microphones,* 2d Ed. Blue Ridge Summit, PA: TAB Books, 1977.

Cooper, Jeff. *Building a Recording Studio.* Rev. Ed. Calabasas, CA: Synergy Group, 1984.

Davis, Don & Carolyn. *Sound System Engineering,* 2d Ed. Indianapolis: Howard W. Sams & Co., Inc., 1986.

Eargle, John. *Handbook of Recording Engineering.* New York: Van Nostrand Reinhold Co., 1987.

_____*The Microphone Handbook.* Plainview, NY: Elar Publishing, 1981.

Everard, Chris. *The Home Recording Handbook.* New York: Amsco Publications, 1986.
Everest, F. Alton. *How to Build a Small Budget Recording Studio From Scratch.* Blue Ridge Summit, PA: TAB Books, 1979.
Keene, Sherman. *Practical Techniques for the Recording Engineer.* Hollywood, CA: Sherman Keene Publications, 1981.
Nisbett, Alec. *The Use of Microphones,* 2d Ed. New York: Hastings House, 1977.
Pohlmann, Ken. *Principles of Digital Audio,* Second Edition. Indianapolis: Howard W. Sams & Co., Inc., 1989.
Rona, Jeff. *Midi—In's, Out's, and Thru's.* Milwaukee: Hal Leonard Publishing Co., 1987.
Ruggeberg, Rand. *1988 Songwriter's Market.* Cincinnati-Writer's Digest Books.
Runstein, Robert and Huber, David. *Modern Recording Techniques,* 2d Ed. Indianapolis: Howard W. Sams & Co., Inc., 1986.
Williams, George. *The Songwriter's Demo Manual and Success Guide.* Bridgeport, CA: Tree by the River, 1984.
Woram, John. *Sound Recording Handbook.* Indianapolis: Howard W. Sams & Co., Inc., 1989.

The following books are not in the Mix Bookshelf Catalog, but they are also useful references.

Are You Ready for Multitrack? Montebello, CA: Teac Corp.
Connelly, Will. *The Musician's Guide to Independent Record Production.* Chicago: Contemporary Books, Inc.
Everest, F. Alton. *Handbook of Multichannel Recording.* Blue Ridge Summit, PA: TAB Books, 1975.
Hickman, Walter A. *Time Code Handbook.* Frederick, MD: Cipher Digital Inc., 1987.
Martin, George. *All You Need Is Ears.* New York: St. Martin's Press, 1979.
Rapaport, Diane Sward. *How to Make and Sell Your Own Record.* Tiburon, CA: Headland Press, 1984.
Rosmini, Dick. *Teac Multitrack Primer.* Montebello, CA: Teac Corp., 1978.

These recording industry magazines were helpful.

db, 203 Commack Rd., Suite 1010, Commack, NY 11725.
Home & Studio Recording, Music Maker Publications, Inc., 22024 Lassen St., Suite 118, Chatsworth, CA 91311.

Mix, Mix Publications, 6400 Hollis St., #12, Emeryville, CA 94608.

Modern Recording & Music, now part of *db* magazine.

Music & Sound Output, Testa Communications, 25 Willowdale Ave., Port Washington, NY 11050.

Recording Engineer/Producer, 8330 Allison Ave., Suite C, La Mesa, CA 92041.

Guides, Brochures, and Other Literature

The Mix Bookshelf Catalog, mentioned earlier, describes many excellent books on recording techniques, audio, studio construction, microphones, MIDI, and the music business. Also available are *Careers in Audio Engineering* and the *Journal of the Audio Engineering Society,* from the Audio Engineering Society, 60 E. 42nd St., New York, NY 10165.

Microphone application guides are available from Crown International, 1718 W. Mishawaka Rd., Elkhart, IN 46517; Shure Brothers Inc., 222 Hartrey Ave., Evanston, IL 60204; Countryman Associates Inc., 417 Stanford Ave., Redwood City, CA 94063; AKG Acoustics Inc., 77 Selleck St., Stamford, CT 06902; Sennheiser Electronic Corp., 6 Vista Drive, P.O. Box 987, Old Lyme, CT 06371; and Audio-Technica U.S. Inc., 1221 Commerce Drive, Stow, Ohio 44224.

You can find much valuable information in operation manuals and free descriptive literature provided by manufacturers of recording equipment.

The International MIDI Association (11857 Hartsook St., N. Hollywood, CA 91607) sells MIDI technical information, such as the MIDI 1.0 specification and a 50-page detailed explanation of MIDI.

Recording Schools

Each July issue of *Mix* magazine contains a comprehensive directory of recording schools, seminars, and programs. Universities and colleges in most major cities have recording-engineering courses. Listed here are some of the better-known schools. Investigate them thoroughly, however, before making a decision.

The Aspen Audio Recording Institute, Box AA, Aspen, CO 81612. (303) 925-3254

Berklee College of Music, 1140 Boylston St., Boston, MA 02215. (617) 266-1400

California State University, Dominguez Hills, 1000 E. Victoria St., Carson, CA 90747. (213) 516-3543

Center for the Media Arts, Conservatory of Music for the Media, 226 W. 26th St., New York, NY 10001. (212) 807-6670

College for Recording Arts, 665 Harrison St., San Francisco, CA 94107. (415) 781-6306

Fullerton College, Music Dept., 321 E. Chapman Ave., Fullerton, CA 92634. (714) 871-8000 ext. 336

Full Sail Center for the Recording Arts, 660 Douglas Ave., Altamonte Springs, FL 32714. (800) 221-2747, in Florida (305) 788-2450

Georgia State University, Dept. of Commercial Music/Recording of the College of Public and Urban Affairs, University Plaza, Atlanta, GA 30303. (404) 658-3513

Golden West Community College, 15744 Golden West St., Huntington Beach, CA 92647. (714) 895-8780

Dick Grove School of Music, 12754 Ventura Blvd., Studio City, CA 91604. (818) 985-0904

Houston Community College System, 901 Yorkchester, Houston, TX 77079. (713) 468-6891

Indiana University School of Music, Bloomington, IN 47405. (812) 335-1613, 335-1900

Institute of Audio Research, 64 University Pl., New York, NY 10003. (212) 677-7580

Institute of Audio-Video Engineering, 1831 Hyperion Ave., Dept. E, Hollywood, CA 90027. (213) 666-3003 ext. 6

ITM Workshop of Recording Arts, Box 686, Knox, PA 16232. (814) 797-5883

Los Angeles Recording Workshop, 5287 Sunset Blvd., Hollywood, CA 90027. (213) 465-4254

Loyola Marymount University, Dept. of Communication Arts, Loyola Blvd. at W. 80th St., Los Angeles, CA 90045. (213) 642-3033.

McGill University, Faculty of Music, 555 Sherbrooks St. W., Strathcona Music Bldg., Montreal, P.Q., Canada H3A 1E3. (514) 392-5776

McLennan Community College, Commercial Music Dept., 1400 College Dr., Waco, TX 76708. (817) 756-6551

Memphis State University, Music Dept., Memphis, TN 38152. (901) 454-2559

The Music Business Institute, 3376 Peachtree Rd., Atlanta, GA 30326. (800) 554-3346, (404) 231-3303

Northeast Technical Community College, 801 E. Benjamin Ave., P.O. Box 469, Norfolk, NE 68701. (402) 371-2020

The Omega Studio's School of Applied Recording Arts and Sciences, Omega Recording Studios, 5609 Fishers Ln., Rockville, MD 20852. (301) 946-4686

Peabody Institute of the Johns Hopkins University, 1 E. Mt. Vernon Pl., Baltimore, MD 21202. (301) 659-8136

The Recording Workshop, 455 Massieville Rd., Chillicothe, OH 45601. (800) 848-9900, (614) 663-2544

School of Audio Engineering, Suite 110, 3000 S. Robertson Blvd., Los Angeles, CA 90034. (213) 559-0973

SKE Publishing, P.O. Box 2519-M, Sedona, AZ 86336. (602) 282-1258

Sound Master Recording Engineer Schools, Audio-Video Institute, 10747 Magnolia Blvd., North Hollywood, CA 91601. (213) 650-8000

South Plains College, 1401 College Ave., Levelland, TX 79336. (806) 894-9611 ext. 271

State University of New York, Fredonia, Mason Hall, Fredonia, NY 14063. (716) 673-3221

Studio Production Techniques, P.O. Box 741444, Dallas, TX 75374. (214) 426-3766

Texarkana College Recording Studios, 2500 N. Robinson Rd., Texarkana, TX 75501. (214) 838-4541 ext. 257 or 360

Trebas Institute of Recording Arts, 6602 Sunset Blvd., Los Angeles, CA 90028. (213) 467-6800

University of Lowell, College of Music, One University Ave., Lowell, MA 01854. (617) 452-5000

University of Miami, School of Music, Coral Gables, FL 33124 (305) 284-2439

Glossary

A-B—A listening comparison between two audio programs, or between two components playing the same program, performed by switching immediately from one to the other. The levels of the two signals are matched. *See also* Spaced Pair.

Accent Microphone—*See* Spot Microphone.

Access Jacks—Two jacks in a console input module or output module that allow access to points in the signal path, usually for connecting a compressor. Plugging into the access jacks breaks the signal flow and allows you to insert a signal processor in series with the signal.

Alignment—The adjustment of tapehead azimuth and of tape-recorder circuitry to achieve optimum performance from the particular type of tape being used.

Alignment Tape—A pre-recorded tape with tones for alignment of a tape recorder.

Ambience—Room acoustics, early reflections and reverberation. Also, the audible sense of a room or environment surrounding a recorded instrument.

Ambience Microphone—A microphone placed relatively far from its sound source to pick up ambience.

Amplitude, Peak—*See* Peak Amplitude.

Analog-to-Digital (A/D) Converter—A circuit that converts an analog audio signal into a digital bit stream.

Assign—To route or send an audio signal to one or more selected channels.

Attack—The beginning of a note. The first portion of a note's envelope in which a note rises from silence to its maximum volume.

Attack Time—In a compressor, the time it takes for gain reduction to occur in response to a musical attack.

Attenuate—To reduce the level of a signal.

Attenuator—In a mixer or mixing console input module, an adjustable resistive network that reduces the microphone signal level to prevent overloading of the input transformer and mic preamplifier.

Automated Mixing—A system of mixing in which a computer remembers and updates console settings so that a mix can be performed and refined in several stages.

Auxiliary Bus (Aux Bus)—*See* Effects Bus.

Auxiliary Send (Aux Send)—*See* Echo Send.

A-Weighting—*See* Weighted.

Azimuth—In a tape recorder, the angular relationship between the head gap and the tape path.

Azimuth Alignment—The mechanical adjustment of the record or playback head to bring it into proper alignment (90°) with the tape path.

Back-Timing—A technique of cueing up the musical background to a narration track so that the music ends simultaneously with the narration.

Balance—The relative volume levels of various tracks or instruments.

Balanced Line—A cable with two conductors surrounded by a shield, in which each conductor is at equal impedance to ground. With respect to ground, the conductors are at equal potential but opposite polarity.

Bandpass Filter—In a crossover network, a filter that passes a band or range of frequencies but sharply attenuates or rejects frequencies outside the band.

Basic Tracks—Recorded tracks of rhythm instruments (bass, guitar, drums, and, sometimes, the keyboard).

Bass Trap—An assembly that absorbs low-frequency sound waves.

Bi-Amplification (Bi-Amping)—Driving a woofer and tweeter with separate power amplifiers. An active crossover is connected ahead of these power amplifiers.

Bias—In tape-recorder electronics, an ultrasonic signal that drives the erase head, and also is mixed with the audio signal applied to the record head to reduce distortion.

Bidirectional Microphone—A microphone that is most sensitive to sounds arriving from two directions—in front of and behind the microphone. It rejects sounds approaching either side of the microphone. Also called a cosine or figure-eight microphone because of the shape of its polar pattern.

Binaural Recording—A 2-channel recording made with an omnidirectional microphone in each ear of a human or a dummy head, for playback over headphones. The object is to duplicate the acoustic signal appearing at each ear.

Blumlein Array—A stereo microphone technique in which two coincident bidirectional microphones are angled 90° apart (45° to the left and right of center).

Board—*See* Mixing Console.

Bouncing Tracks—A process in which two or more tracks are mixed, and the mixed tracks are recorded on an unused track. Then the original tracks can be erased, which frees them up for recording more instruments.

Boundary Microphone—A microphone designed to be used on a boundary (a hard reflective surface). The microphone capsule is mounted very close to the boundary so that direct and reflected sounds arrive at the microphone diaphragm in phase (or nearly so) at all frequencies in the audible band.

Breathing—The unwanted audible rise and fall of background noise that may occur with a compressor. Also called *pumping*.

Bulk Tape Eraser—A large electromagnet used to erase a whole reel of recording tape at once.

Bus—A common connection of many different signals. An output of a mixer or submixer. A channel that feeds a tape track, signal processor, or power amplifier.

Bus Master—A potentiometer (fader or volume control) in the output section of a mixing console that controls the output level of a bus.

Bus Trim—A control in the output section of a mixing console that provides variable gain control of a bus, used in addition to the bus master for fine adjustment.

Buzz—An unwanted edgy tone that sometimes accompanies audio, containing upper harmonics of 60 Hz.

Calibration—*See* Alignment.

Capacitor Microphone—*See* Condenser Microphone.

Capstan—In a tape-recorder transport, a rotating post that contacts the tape (along with the pinch roller) and pulls the tape past the heads at a constant speed during recording and playback.

Cardioid Microphone—A unidirectional microphone with side attenuation of 6 dB and maximum rejection of sound at the rear of the microphone (180° off-axis). A microphone with a heart-shaped directional pattern.

Channel—A single path of an audio signal. Usually, each channel contains a different signal.

Channel Assign—*See* Assign.

Chorus—A special effect in which a signal is delayed by 15 to 35 milliseconds, where the delayed signal is combined with the original signal, and the delay is varied randomly or periodically. This creates a wavy, multiple-voice effect. Sometimes a portion of the output signal is fed back into the input. Also, the main portion of a song that is repeated several times throughout the song with the same lyrics.

Clean—Free of noise, distortion, overhang, leakage. Not muddy.

Clear—Easy to hear, easy to differentiate. Reproduced with sufficient high frequencies.

Close-Field Monitoring—A monitor-speaker arrangement in which the speakers are placed very near the listener (usually on top of the console meter bridge) to reduce the audibility of control-room acoustics.

Coincident Pair—A stereo microphone, or two separate microphones, placed so that the microphone diaphragms occupy approximately the same point in space. They are angled apart and mounted one directly above the other.

Comb-Filter Effect—The frequency response caused by combining a sound with its delayed replica. The frequency response has a series of peaks and dips caused by phase interference. The peaks and dips resemble the teeth of a comb.

Combining Amplifier—An amplifier at which the outputs of two or more signal paths are mixed together, to feed a single track of a tape recorder.

Combining Network—A resistive network at which the outputs of two or more signal paths are mixed together, to feed a single track of a tape recorder.

Complex Wave—A wave with more than one frequency component.

Compression—The portion of a sound wave in which molecules are pushed together, forming a region with higher-than-normal atmospheric pressure. Also, in signal processing, the reduction in dynamic range caused by a compressor.

Compression Ratio (Slope)—In a compressor, the ratio of the change in input level (in dB) to the change in output level (in dB). For example, a 2:1 ratio means that for every 2-dB change in input level, the output level changes 1 dB.

Compressor—A signal processor that reduces dynamic range by means of automatic volume control. An amplifier whose gain decreases as the input signal level increases above a preset point.

Condenser Microphone—A microphone that works on the principle of variable capacitance to generate an electrical signal. The microphone diaphragm and an adjacent metallic disk (called a backplate) are charged to form two plates of a capacitor. Incoming sound waves vibrate the diaphragm, varying its spacing to the backplate, which varies the capacitance, which varies the voltage between the diaphragm and the backplate.

Connector—A device that makes electrical contact between a signal-carrying cable and an electronic device, or between two cables. A device used to connect or hold together a cable and an electronic component so that a signal can flow from one to the other.

Console—*See* Mixing Console.

Contact Pickup—A transducer that contacts a musical instrument and converts its mechanical vibrations into a corresponding electrical signal.

Control Room—The room in which the engineer controls and monitors the recording.

Crossover—An electronic network that divides an incoming signal into two or more frequency bands.

Crossover, Active (Electronic Crossover)—A crossover network with amplifying components; used ahead of the power amplifiers in a bi-amped or tri-amped speaker system.

Crossover Frequency—The single frequency at which both filters of a crossover network are down 3 dB.

Crossover, Passive—A crossover with passive (nonamplifying) components; used after the power amplifier.

Crosstalk—The unwanted transfer of a signal from one channel to another. Crosstalk often occurs between adjacent tracks within a record or playback head in a tape recorder.

Cue or Cue Send—In a mixing-console input module, a control that adjusts the level of the signal feeding the cue mixer which feeds a signal to headphones in the studio.

Cue Mixer—A submixer in a mixing console that takes signals from cue sends as inputs and mixes them into a composite signal that drives headphones in the studio.

Cue Sheet—Used during mixdown, a chronological list of mixing-console control adjustments required at various points in the recorded song. These points may be indicated by tape-counter or elapsed-time readings.

Cue System—A monitor system that allows musicians to hear themselves and previously recorded tracks through headphones.

dB—Abbreviation for decibel.

dBA—Refers to decibels, A-weighted (*see* Weighted)

dBm—Decibels relative to 1 milliwatt.

dBu—Decibels relative to 0.775 volt.

dBV—Decibels relative to 1 volt.

Dead—Having very little or no reverberation.

Decay—The portion of the envelope of a note in which the envelope goes from maximum to some mid-range level. Also, the decline in level of reverberation over time.

Decay Time—Reverberation time (RT_{60}). The time it takes for reverberation to decay to 60 dB below the original steady-state level.

Decibel—The unit of measurement of audio level. Abbreviated dB.

Decoded Tape—A tape that is expanded after being compressed by a noise-reduction system. Such a program has normal dynamic range.

De-esser—A signal processor that removes excessive sibilance ("s" and "sh" sounds) by compressing high frequencies around 5 to 10 kHz.

Degausser—*See* Demagnetizer.

Delay—The time interval between a signal and its repetition. A digital delay or a delay line is a signal processor that delays a signal for a short time.

Demagnetizer—An electromagnet with a probe tip that is touched to elements of the tape path (such as tape heads and tape guides) to remove residual magnetism.

Depth—The audible sense of nearness and farness of various instruments. Instruments recorded with a high ratio of direct-to-reverberant sound are perceived as being close; instruments recorded with a low ratio of direct-to-reverberant sound are perceived as being distant.

Designation Strip—A strip of paper taped near console faders to designate the instrument that each fader controls.

Design Center—The portion of fader travel (usually shaded), about 10 to 15 dB from the top, in which console gain is distributed for optimum headroom and signal-to-noise ratio. During normal operation, each fader in use should be placed at or near design center.

Desk—British term for mixing console.

Diffusion—An even distribution of sound in a room.

Digital Audio—An encoding of an analog audio signal in the form of binary digits (ones and zeros).

Digital Recording—A recording system in which the audio signal is stored in the form of binary digits.

Digital-to-Analog Converter—A circuit that converts a digital audio signal into an analog audio signal.

Dim—To temporarily reduce the monitor volume by a preset amount (so you can carry on a conversation).

Direct Box—A device used for connecting an amplified instrument directly to a mixer mic input. The direct box converts a high-impedance unbalanced audio signal into a low-impedance balanced audio signal.

Direct Injection (DI)—Recording with a direct box.

Direct Output, Direct Out—Following a mic preamplifier, an output connector which is used to feed the signal of one instrument to one track of a tape recorder.

Direct Sound—Sound traveling directly from the sound source to the microphone (or to the listener) without reflections.

Directional Microphone—A microphone that has different sensitivity in different directions. A unidirectional or bidirectional microphone.

Distortion—An unwanted change in the audio waveform, causing a raspy or gritty sound quality. The appearance of frequencies in a device's output signal that were not in the input signal.

Dolby Tone—A reference tone recorded at the beginning of a Dolby-encoded tape for alignment purposes.

Doubling—A special effect in which a signal is combined with its 15- to 35-millisecond delayed replica. This process mimics the sound of two identical voices or instruments playing in unison.

Drop-Frame—For color video production, a mode of SMPTE time code which causes the time code to match the clock on the wall. Once every minute, except for the tenth minute, frame numbers 00 and 01 are dropped.

Drop-Out—During playback of a tape recording, a momentary loss of high frequencies caused by separation of the tape from the playback head due to dust, tape-oxide irregularity, etc.

Drum Machine—A device that plays memory-chip recordings of real drums.

Dry—Having no echo or reverberation. Referring to a close-sounding signal that is not yet processed by a reverberation or delay device.

Dynamic Microphone—A microphone that generates electricity when sound waves cause a conductor to vibrate in a stationary magnetic field. The two types of dynamic microphone are the moving coil and the ribbon. A moving-coil microphone is usually called a dynamic microphone.

Dynamic Range—The range of volume levels in a program from softest to loudest.

Earth Ground—A connection to moist dirt (the ground we walk on). This connection is usually done via a long copper rod or a cold-water pipe.

Echo—A delayed repetition of a signal or sound. A sound delayed 50 milliseconds or more that is combined with the original sound.

Echo Chamber—A hard-surfaced room containing a widely separated loudspeaker and microphone, used for creating reverberation.

Editing—The cutting and rejoining of magnetic tape to delete unwanted material, to insert leader tape, or to rearrange recorded material into the desired sequence.

Editing Block—A metal block that holds magnetic tape during the editing-splicing procedure.

Effects—Interesting sound phenomena created by signal processors, such as reverberation, echo, flanging, doubling, or chorus.

Effects Bus—The bus that feeds effects devices (signal processors).

Effects Mixer—A submixer in a mixing console that combines signals from effects sends (aux sends) and feeds the mixed signal to the input of a special-effects device, such as a reverberation unit.

Effects Return—In the output section of a mixing console, a control that adjusts the amount of signal received from a reverberation or echo device. The effects-return signal is mixed with the program bus signal.

Effects Send—In an input module of a mixing console, a control that adjusts the amount of signal sent to a special-effects device, such as a reverberation or delay unit. The effects-send control normally adjusts the amount of reverberation or echo heard on each instrument.

Efficiency—In a loudspeaker, the ratio of acoustic power output to electrical power input.

EIA—Electronic Industries Association.

EIA Rating—A microphone-sensitivity specification that states the microphone output level in dBm into a matched load for a given Sound Pressure Level (SPL). SPL + dB (EIA rating) = dBm output into a matched load.

Electret Microphone—A condenser microphone in which the electrostatic field of the capacitor is generated by an electret–a material that permanently stores an electrostatic charge.

Electrostatic Field—The force field between two conductors charged with static electricity.

Electrostatic Interference—The unwanted presence of an electrostatic hum field in signal conductors.

Encoded Tape—A tape containing a signal compressed by a noise-reduction unit.

Envelope—The rise and fall in volume of one note. The envelope connects successive peaks of the waves comprising a note. Each harmonic in the note might have a different envelope.

Equalization (EQ)—The adjustment of frequency response in order to alter the tonal balance or to attenuate unwanted frequencies.

Equalizer—A circuit (usually in each input module of a mixing console, or in a separate unit) that alters the frequency spectrum of a signal passed through it.

Erase—To remove an audio signal from magnetic tape by applying an ultrasonic varying magnetic field so as to randomize the magnetization of the magnetic particles on the tape.

Erase Head—A head in a tape recorder that erases the signal on tape.

Expander—A signal processor that increases the dynamic range of a signal passed through it. An amplifier whose gain decreases as its input level decreases. When used as a noise gate, an expander reduces the gain of low-level signals to reduce noise between notes.

Fade-Out—To gradually reduce the volume of the last several seconds of a recorded song, from full level down to silence, by slowly pulling down the master fader.

Fader—A linear or sliding potentiometer (volume control) that is used to adjust signal level.

Feed—To send an audio signal to some device or system. Also, a feed is an output signal sent to some device or system.

Feedback—The return of some portion of an output signal to the system's input.

Feed Reel—The left-side reel on a tape recorder that unwinds during recording or playback.

Filter—A circuit that sharply attenuates frequencies above or below a certain frequency. Used to reduce noise and leakage above or below the frequency range of an instrument or voice.

Flanging—A special effect in which a signal is combined with its delayed replica, and the delay is varied between 0 and 20 milliseconds. A hollow, swishing, ethereal effect—like a variable-length pipe, or like a jet plane passing overhead. A variable comb filter produces the flanging effect.

Fletcher-Munson Effect—Named after the two people who discovered it, the psychoacoustical phenomenon in which the subjective frequency response of the ear changes with program level. Due to this effect, a program played at a lower volume than the original level subjectively loses low- and high-frequency response.

Float—To disconnect from ground.

Flutter—A rapid periodic variation in tape speed.

Flutter Echoes—A rapid series of echoes that occurs between two parallel walls.

Flux—Magnetic lines of force.

Fluxivity—The measure of the flux density of a magnetic recording tape, per unit of track width.

Fly-In—To copy part of a recorded track on another recorder, and rerecord that copy back to the original multitrack tape in a different part of the song, in sync with other recorded tracks. Example: Copy the vocal track from the first chorus of the song to an external recorder. Rerecord (fly-in) that copy onto the multitrack tape at the second chorus. Then the first and second choruses will have identical vocal performances.

Foldback (FB)—*See* Cue System.

Frequency—The number of cycles per second of a sound wave or an audio signal, measured in Hz. A low frequency (say, 100 Hz) has a low pitch; a high frequency (say, 10,000 Hz) has a high pitch.

Frequency Response—The range of frequencies that an audio device will reproduce at an equal level (within a tolerance, such as ±3 dB).

Full Track—A single tape track recorded across the full width of a tape.

Fundamental—The lowest frequency in a complex wave.

Gain—Amplification. The ratio, expressed in decibels, between the output voltage and the input voltage, or between the output power and the input power.

Gap—In a tape-recorder head, the thin break in the electromagnet that contacts the tape.

Gate—To turn off a signal when its amplitude falls below a preset value. The signal-processing device used for this purpose. *See* Noise Gate.

Generation—A copy of a tape. A copy of the original master recording is a first generation tape. A copy made from the first generation tape is a second generation, and so on.

Generation Loss—The degradation of signal quality (the increase in noise and distortion) that occurs with each successive generation of a tape recording.

Gobo—A movable partition used to prevent the sound of an instrument from reaching another instrument's microphone. Short for "go-between."

Graphic Equalizer—An equalizer with a horizontal row of faders; the fader-knob positions graphically indicate the frequency response of the equalizer. Usually used to equalize monitor speakers for the room they are located in.

Ground—The zero-signal reference point for a system of audio components.

Ground Bus—A common connection to which equipment is grounded, usually a heavy copper plate.

Ground Loop—A loop or circuit formed of ground leads. The loop formed when unbalanced components are connected via two ground paths—the connecting-cable shield and the power ground. Ground loops cause hum and should be avoided.

Grounding—Connecting pieces of electronic equipment to ground. Proper grounding ensures that there is no voltage difference between equipment chassis. An electrostatic shield needs to be grounded to be effective.

Group—*See* Submix.

Guard Band—The spacing between tracks on a multitrack tape or tape head; used to prevent crosstalk.

Half-Track—A tape track recorded across approximately half the width of a tape. A half-track recorder usually records two such tracks simultaneously in the same direction to make a stereo recording.

Harmonic—An overtone whose frequency is a whole-number multiple of the fundamental frequency.

Head—An electromagnet in a tape recorder that either erases the audio signal on tape, records a signal on tape, or plays back a signal that is already on tape.

Head Gap—*See* Gap.

Headphones—A head-worn transducer that covers the ears and converts electrical audio signals into sound waves.

Headroom—The safety margin, measured in decibels, between the signal level and the maximum undistorted signal level. In a tape recorder, the dB difference between standard operating level (corresponding to a 0-VU reading) and the level causing 3% total harmonic distortion. High-frequency headroom increases with tape speed.

Hertz (Hz)—Cycles per second, the unit of measurement of frequency.

High-Pass Filter—A filter that passes frequencies above a certain frequency and attenuates frequencies below that same frequency. A low-cut filter.

Hiss—A noise signal containing all frequencies, but with greater energy at higher octaves. Hiss sounds like wind blowing through trees. It is usually caused by random signals generated by microphones, electronics, and magnetic tape.

Hot—A high recording level causing slight distortion, used for special effect. Also, a condition in which a chassis or circuit has a potentially dangerous voltage on it. Also, referring to the conductor in a microphone cable which has a positive voltage on it at the instant that sound pressure moves the diaphragm inward.

Hum—An unwanted low-pitched tone (60 Hz and its harmonics) heard along with the audio signal.

Hypercardioid Microphone—A directional microphone with a polar pattern that has 12 dB attenuation at the sides, 6 dB attenuation at the rear, and two nulls of maximum rejection at 110° off axis.

Hz—Abbreviation for hertz.

Image—An illusory sound source located between two stereo speakers.

Impedance—The opposition of a circuit to the flow of alternating current. Impedance, abbreviated Z, is the complex sum of resistance and reactance.

Input—The connection going into an audio device. In a mixer or mixing console, a connector for a microphone or other signal source.

Input Attenuator—*See* Attenuator.

Input Module—In a mixing console, the set of controls affecting a single input signal. An input module usually includes an attenuator, fader, equalizer, aux sends, and channel-assign controls.

Input Section—The row of input modules in a mixing console.

Input/Output (I/O) Console (In-Line Console)—A mixing console arranged so that input and output sections are vertically aligned. Each module (other than the monitor section) contains one input channel and one output channel.

Jack—A female or receptacle-type connector for audio signals into which a plug is inserted.

Kilo—A prefix meaning one thousand. Abbreviated k.

Lay-In—*See* Fly-In.

Leadering—The process of splicing leader tape between program selections.

Leader Tape—Plastic or paper tape without an oxide coating that is used for a spacer between takes (i.e., for silence between songs).

Leakage—The overlap of an instrument's sound into another instrument's microphone. Also called *bleed* or *spill*.

LEDE—Abbreviation for Live-End Dead-End; a type of control room acoustic treatment in which the front half of the control room prevents early reflections to the mixing position, while the back half of the control room reflects diffused sound to the mixing position.

LED Indicator—A recording-level indicator using one or more light emitting diodes.

Level—The degree of intensity of an audio signal—the voltage, power, or sound pressure level. The original definition of level is the power in watts.

Level Setting—In a tape recorder, the process of adjusting the level of the signal sent to the record head so that maximum tape magnetization occurs without distortion. A VU meter or other indicator shows recording level.

Limiter—A signal processor whose output is constant above a preset input level. A compressor with a compression ratio of 10:1 or greater, with the threshold set just below the point of distortion of the following device. Used to prevent distortion of attack transients or peaks.

Line Level—In balanced professional recording equipment, a signal whose level is approximately 1.23 volts (+4 dBm). In unbalanced equipment (most home hi-fi or semi-pro recording equipment), a signal whose level is approximately 0.316 volt (−10 dBV).

Live—Having audible reverberation. Also, occurring in real time, in person.

Live Recording—A recording made at a concert. Also, a recording made of a musical ensemble playing all at once, rather than overdubbing.

Localization—The ability of the human hearing system to tell the direction of a real or illusionary sound source.

Loudspeaker—A transducer that converts electrical energy (the signal) into acoustical energy (sound waves).

Lowpass Filter—A filter that passes frequencies below a certain frequency and attenuates frequencies above that same frequency. A high-cut filter.

Magnetic Recording Tape—A recording medium made of magnetic particles (usually ferric oxide) suspended in a binder and coated on a long strip of thin plastic (usually Mylar).

Mask—To hide or cover up one sound with another sound. To make a sound inaudible by playing another sound along with it.

Master Fader—A volume control that affects the level of all program buses simultaneously.

Master Tape—A completed tape used to generate tape copies or disks.

Memory—A group of integrated circuit chips, each containing thousands of solid-state components, used to temporarily or permanently store digital data (such as an audio signal in digital format).

Memory Rewind—A tape-recorder function that rewinds the tape to a preset tape-counter position.

Meter—A device that indicates voltage, resistance, current, or signal level.

Mic—An abbreviation for microphone.

Mic Level—The level or voltage of a signal produced by a microphone, typically 2 millivolts.

Microphone—A transducer or device that converts an acoustical signal (sound) into a corresponding electrical signal.

Microphone Techniques—The selection and placement of microphones to pick up sound sources.

MIDI—Abbreviation for Musical Instrument Digital Interface, a specification for a connection between synthesizers, drum machines, and computers that allows them to communicate with and/or control each other.

MIDI Channel—A route for the transmission and reception of MIDI signals. Each channel controls a separate MIDI musical instrument. Up to 16 channels can be sent on a single MIDI cable.

MIDI In—A connector in a MIDI device that receives MIDI signals.

MIDI Out—A connector in a MIDI device that transmits MIDI signals.

MIDI Thru—A connector in a MIDI device that duplicates the MIDI information at the MIDI-in connector. Used to route the MIDI-In signal to another MIDI device in series.

Mid-Side—A coincident-pair stereo microphone technique using a forward-facing unidirectional, omnidirectional, or bidirectional microphone and a side-facing bidirectional microphone. The microphone signals are summed and differenced to produce right- and left-channel signals.

Mike—To pick up with a microphone. Also, abbreviation for microphone (used by older radio operators).

Milli—A prefix meaning one thousandth, abbreviated m.

Mix—To combine two or more different signals into a common signal. Also, a control on a delay unit that varies the ratio between the dry signal and the delayed signal.

Mixdown—The process of playing prerecorded tape tracks through a mixing console and mixing them to two stereo channels for recording on a 2-track tape recorder.

Mixer—A device that mixes or combines audio signals and controls the relative levels of the signals.

Mixing Console—A large mixer with additional functions such as equalization or tone control, pan pots, monitoring controls, channel assigns, and control of signals sent to external signal processors.

Monaural—Referring to listening with one ear. Often incorrectly used to mean monophonic.

Monitor—To listen to an audio signal with headphones or a loudspeaker.

A monitor is a loudspeaker in a control room used for monitoring.

Monitoring—Listening to an audio signal with a monitor.

Mono-Compatible—A characteristic of a stereo program, in which the program channels can be combined to a mono program without altering the frequency response or balance. A mono-compatible stereo program has the same frequency response in either stereo or mono because there is no delay or phase shift between channels to cause phase interference.

Monophonic (Mono)—Refers to a single channel of audio. A monophonic program can be played over one or more loudspeakers or one or more headphones. Also, describing a synthesizer that plays only one note at a time (not chords).

Moving-Coil Microphone—A dynamic microphone in which the conductor is a coil of wire moving in a fixed magnetic field. The coil is attached to a diaphragm which vibrates when struck with sound waves.

M-S Recording—*See* Mid-side.

Muddy—Unclear sounding; having excessive leakage, reverberation, or overhang.

Multiple-D Microphone—A directional microphone which has multiple sound-path lengths between its front and rear sound entries. This type of microphone has minimal proximity effect.

Multiprocessor—A signal processor that can perform several different signal-processing functions.

Multi-timbral—In a synthesizer, the ability to produce two or more different voices (patches or timbres) at the same time.

Multitrack—Having more than two tape tracks when referring to a tape recorder or tape-recorder head.

Mute—To turn off an input signal on a mixing console by disconnecting the input-module output from channel assign and direct out. The mute function is used to reduce tape noise during silent portions of tracks.

Near-Coincident—A stereo microphone technique in which the two microphones are angled apart symmetrically on either side of center and horizontally spaced a few inches apart.

Noise—Unwanted sound, such as hiss from electronics or tape. An audio signal with an irregular, nonperiodic waveform.

Noise-Reduction System—A signal processor used to reduce tape hiss (and sometimes print-through) caused by the recording process. Some of these systems compress the signal during recording and expand it in a complementary fashion during playback.

Noise Gate—A gate used to reduce or eliminate noise between notes.

Octave—The interval between any two frequencies where the upper frequency is twice the lower frequency.

Off-Axis—Not directly in front of a microphone or loudspeaker.

Off-Axis Coloration—In a microphone, the deviation from the on-axis frequency response that sometimes occurs at angles off the axis of the microphone. The coloration of sound (alteration of tone quality) for sounds arriving off-axis to the microphone.

Omnidirectional Microphone—A microphone that is equally sensitive to sounds arriving from all directions.

On-Location Recording—A recording made outside the studio, in a room or hall where the music is normally performed or practiced.

Open Tracks—On a multitrack tape recorder, tracks that have not yet been used.

Outboard Equipment—Signal processors that are external to the mixing console.

Output—A connector in an audio device from which the signal comes, and feeds successive devices.

Out-Take—A take, or section of a take, that is to be removed or not used.

Overdub—To record a new musical part on an unused track in synchronization with previously recorded tracks.

Overhang—The continuation of a signal at the output of a device after the input signal has ceased.

Overload—The distortion that occurs when an applied signal exceeds a system's maximum output level.

Overtone—A frequency component of a complex wave which is higher than the fundamental frequency.

Pad—*See* Attenuator.

Pan Pot—Abbreviation for panoramic potentiometer. In each input module in a mixing console, a control that divides a signal between two channels in an adjustable ratio. By doing so, a pan pot controls the location of a sonic image between a stereo pair of loudspeakers.

Parametric Equalizer—An equalizer with continuously variable parameters, such as frequency, bandwidth, and amount of boost or cut.

Patch—To connect one piece of audio equipment to another with a cable. Also, a setting of synthesizer parameters to achieve a sound with a certain timbre.

Patch Bay (Patch Panel)—An array of connectors, usually in a rack, to which equipment inputs and outputs are wired. A patch bay makes it easy to interconnect various pieces of equipment in a central, accessible location.

Patch Cord—A short length of cable with a coaxial plug on each end, used for signal routing in a patch bay.

Peak—On a graph of a sound wave or signal, the highest point in the waveform. The point of greatest voltage or sound pressure in a cycle.

Peak Amplitude—On a graph of a sound wave, the sound pressure of the waveform peak. On a graph of an electrical signal, the voltage of the waveform peak. The amplitude of a sound wave or signal as measured on a meter is 0.707 times the peak amplitude.

Peak Program Meter (PPM)—A meter that responds fast enough to closely follow the peak levels in a program.

Peaking Equalizer—An equalizer that provides maximum cut or boost at one frequency, so that the resulting frequency response of a boost resembles a mountain peak.

Period—The time between the peak of one wave and the peak of the next. The time between corresponding points on successive waves. Period is the inverse of frequency.

Personal Studio—A minimal group of recording equipment set up for one's personal use, usually using a 4-track cassette recorder-mixer. Also, a simple 4-track cassette recorder-mixer for one's personal use.

Perspective—In the reproduction of a recording, the audible sense of distance to the musical ensemble, the point of view. A close perspective has a high ratio of direct sound to reverberant sound; a distant perspective has a low ratio of direct sound to reverberant sound.

PFL—Abbreviation for Pre-Fader Listen. *See* Solo.

Phantom Power—A dc voltage (usually 12 to 48 volts) that is applied to microphone signal conductors to power condenser microphones.

Phase—The degree of progression in the cycle of a wave, where one complete cycle is 360°.

Phase Cancellation, Phase Interference—The cancellation of certain frequency components of a signal that occurs when the signal is combined with its delayed replica. At certain frequencies, the direct and delayed signals are of equal level and opposite polarity (180° out of phase), and when combined, they cancel out. The result is a comb-filter frequency response having a periodic series of peaks and dips. Phase interference can occur between the signals of two microphones picking up the same source at different distances, or can occur at a microphone picking up both a direct sound and its reflection from a nearby surface.

Phase Shift—The difference in degrees of phase angle between corresponding points on two waves. If one wave is delayed with respect to another, there is a phase shift between them of $2\pi ft$, where π = 3.14, f = frequency in Hz, and t = delay in seconds.

Phasing—A special effect in which a signal is combined with its phase-shifted replica to produce a variable comb-filter effect. *See* also Flanging.

Phono Plug—A cylindrical plug used with headphones, microphones, and other audio equipment (usually ¼-inch diameter). An unbalanced phone plug has a tip for the hot signal and a sleeve for the shield or ground. A balanced phone plug has a tip for the hot signal, a ring for the return signal, and a sleeve for the shield or ground.

Phone Plug—A coaxial plug with a central pin for the hot signal and a ring of pressure-fit tabs for the shield or ground. Also called an RCA plug.

Pickup—A contact pickup. Also, a transducer in an electric guitar that converts string motion to a corresponding electrical signal.

Pinch Roller—In a tape-recorder transport, the rubber wheel that pinches or traps the tape between itself and the capstan, so that the capstan can move the tape.

Ping-Ponging—*See* Bouncing Tracks.

Pink Noise—A noise signal containing all frequencies (unless band-limited), with equal energy per octave. Pink noise is a test signal used for equalizing a sound system to the desired frequency response, and for testing loudspeakers.

Pitch—The subjective lowness or highness of a tone; the pitch of a tone usually correlates with the fundamental frequency.

Pitch Control—A control on a tape recorder that varies the tape speed, thereby varying the pitch of the signal on the tape. The pitch control is used to match the pitch of pre-recorded instruments with that of an instrument to be overdubbed.

Pitch Shifter—A signal processor that changes the pitch of an instrument without changing its duration.

Playback Equalization—In tape-recorder electronics, fixed equalization applied to the signal during playback to compensate for certain losses.

Playback Head—The head in a tape recorder that picks up a prerecorded magnetic signal from the moving tape and converts it to a corresponding electrical signal.

Plug—A male connector that inserts into a jack (a female connector).

Polar Pattern—The directional pickup pattern of a microphone. A graph of microphone sensitivity plotted vs. angle of sound incidence. Examples of polar patterns are omnidirectional, bidirectional, and unidirectional. Subsets of unidirectional are cardioid, supercardioid, and hypercardioid.

Polarity—Referring to the positive or negative direction of an electric, acoustic, or magnetic force. Two identical signals in opposite polarity are 180° out-of-phase with each other at all frequencies.

Polyphonic—Describing a synthesizer that can play more than one note at a time (chords).

Pop—A thump or little explosion sound heard in a vocalist's microphone signal. Pop occurs when the user says words with a "p," "t," or "b" so that a turbulent puff of air is forced from the mouth and strikes the microphone diaphragm.

Pop Filter—A screen placed on a microphone grille that attenuates or filters out pop disturbances before they strike the microphone diaphragm. Usually made of open-cell plastic foam or silk, a pop filter reduces pop and wind noise.

Portable Studio—A combination tape recorder and mixer in one portable chassis.

Post-Echo—A repetition of a sound following the original sound, caused by print-through.

Power Amplifier—An electronic device that amplifies or increases the power level fed into it to a level sufficient to drive a loudspeaker.

Power Ground—A connection to the power company's earth ground through the U-shaped hole in a power outlet. In the power cable of an electronic component with a three-prong plug, the U-shaped prong is wired to the component's chassis. This wire conducts electricity to power ground if the chassis becomes electrically hot, preventing shocks. Also called *safety ground.*

Preamplifier—In an audio system, the first stage of amplification that boosts a mic-level signal to line level.

Predelay—Short for prereverberation delay. The delay (about 30 to 100 milliseconds) between the arrival of the direct sound and the onset of reverberation. Usually, the longer the predelay, the greater the perceived room size.

Pre-Echo—A repetition of a sound that occurs before the sound itself, caused by print-through.

Prefader/Postfader Switch—A switch that selects a signal either ahead of the fader (prefader) or following the fader (postfader). The level of a prefader signal is independent of the fader position; the level of a postfader signal follows the fader position.

Preproduction—Planning in advance what you're going to do at a recording session, in terms of track assignments, overdubbing, studio layout, and microphone selection.

Presence—The audible sense that a reproduced instrument is present in the listening room. Some synonyms are closeness, definition, and punch. Presence is often created by an equalization boost in the midrange or upper midrange.

Pressure Zone Microphone—A boundary microphone constructed with the microphone diaphragm parallel with, and facing, a reflective surface.

Preverb—A special effect in which the reverberation of a note precedes it, rather than follows it. Preverb is achieved by playing an instrument's track backwards while adding reverberation to it, and recording the reverberation on an unused track. When the tape is reversed so that the instrument's track plays forward, preverb is heard as the reverberation plays backwards.

Print-Through—The transfer of a magnetic signal from one layer of tape to the next on a reel, causing an echo preceding or following the program.

Production—A recording that is enhanced by special effects. Also, the supervision of a recording session to create a satisfactory recording. This involves getting musicians together for the session, making musical suggestions to the musicians to enhance their performance, and making suggestions to the engineer for sound balance and effects.

Program Bus—A bus or output that feeds an audio program to a tape-recorder track.

Program Mixer—In a mixing console, a mixer formed of input-module outputs, combining amplifiers, and program buses.

Proximity Effect—The bass boost that occurs with a single-D directional microphone when it is placed a few inches from a sound source. The closer the microphone, the greater the low-frequency boost due to proximity effect.

Punch In/Out—A feature in a tape recorder that lets you insert a recording of a corrected musical part into a previously recorded track by going into and out of record mode as the tape is rolling.

Pure Waveform—A waveform of a single frequency; a sine wave. A pure tone is the perceived sound of such a wave.

Quarter-Track—A tape track recorded across one-quarter of the width of the tape. A quarter-track recorder usually records two stereo programs (one in each direction).

Rack—A 19-inch-wide wooden or metal cabinet used to hold audio equipment.

Radio-Frequency Interference (RFI)—Radio-frequency electromagnetic waves induced in audio cables or equipment, causing various noises in the audio signal.

Rarefaction—The portion of a sound wave in which molecules are spread apart, forming a region with lower-than-normal atmospheric pressure. The opposite of compression.

Real-Time Recording—Recording notes into a sequencer, in the correct tempo, for playback at the same tempo. Also, a recording made direct to lacquer disc or direct to 2-track without any overdubs or mixdown.

Recirculation (Regeneration)—Feeding the output of a delay device back into its input to create multiple echoes. Also, the control on a delay device that affects how much delayed signal is recycled to the input.

Record—To store an event in permanent form. Usually, to store an audio signal in magnetic form on magnetic tape.

Record Equalization—In the tape-recorder electronics, equalization applied to the signal during recording to compensate for certain losses.

Record Head—The head in a tape recorder that puts the audio signal on tape by magnetizing the tape particles in a pattern corresponding to the audio signal.

Recorder-Mixer—A combination tape recorder and mixer in one chassis.

Recording-Reproduction Chain—The series of events and equipment that are involved in sound recording and playback.

Reflected Sound—Sound waves that reach the listener after being reflected from one or more surfaces.

Release—The final portion of a note's envelope in which the note falls from its sustain level back to silence.

Release Time (Recovery Time)—In a compressor, the time it takes for the gain to return to normal after the end of a loud passage.

Remix—To mix again; to do another mixdown with different console settings or different editing.

Remote Recording—*See* On-Location Recording.

Resistance—The opposition of a circuit to the flow of current. Resistance, abbreviated Ω, is measured in ohms and may be calculated by dividing voltage by current.

Resistor—An electronic component that opposes current flow.

Return-to-Zero—*See* Memory Rewind.

Reverberation—The persistence of sound in a room after the original sound has ceased. It is caused by multiple sound reflections (echoes) that decrease in intensity with time, and are so closely spaced in time as to merge into a single continuous sound, which, eventually, is completely absorbed by the inner surfaces of the room. The timing of the echoes is random, and the echoes increase in number as they decay. An example of reverberation is the sound you hear just after you shout in an empty gymnasium. An *echo* is a discrete repetition of a sound, while *reverberation* is a continuous fade-out of sound. *Artificial reverberation* is reverberation in an audio signal created mechanically or electronically rather than acoustically.

Reverberation Time—Abbreviated RT_{60}, the time it takes for reverberation to decay to 60 dB below the original steady-state level. RT_{60} is usually measured at 500 Hz.

Reverse Echo—A multiple echo that precedes the sound that caused it, building up from silence into the original sound. This special effect is created in a manner similar to preverb.

RFI—*See* Radio Frequency Interference.

Rhythm Tracks—The recorded tracks of the rhythm instruments (guitar, bass, drums, and sometimes keyboards).

Ribbon Microphone—A dynamic microphone in which the conductor is a long metallic diaphragm (ribbon) suspended in a magnetic field.

Ride Gain—To turn down the volume of a microphone when the source gets louder, and turn up the volume when the source gets quieter, in an attempt to reduce dynamic range.

Room Modes—*See* Standing Wave.

Safety Copy—A copy of the master tape, to be used if the master tape is lost or damaged.

Safety Ground—*See* Power Ground.

Sampling—Recording a short sound event into computer memory. The audio signal is converted into digital data representing the signal waveform, and the data is stored in memory chips for later playback.

Saturation—Overload of a magnetic tape. The point at which a further increase in magnetizing force does not cause an increase in magnetization of the tape-oxide particles.

Scratch Vocal—A vocal performance that is done simultaneously with the rhythm instruments so that the musicians can keep their place in the song and get a feel for the song. Since it contains leakage, the scratch-vocal recording is usually erased. Then the singer overdubs the vocal part that is to be used in the final recording.

Sensitivity—The output of a microphone in volts for a given input in sound pressure level. Also, the sound pressure level a loudspeaker produces at four feet when driven with one watt of pink noise.

Sequencer—A device that records a series of synthesizer note parameters into computer memory chips for later playback. A computer can act as a sequencer when it runs a sequencer program. During playback, the sequencer activates the synthesizer sound generators.

Shelving Equalizer—An equalizer that applies a constant boost or cut above or below a certain frequency, so that the shape of the frequency response resembles a shelf.

Shield—A conductive enclosure (usually metallic) around one or more signal conductors that is used to keep out electrostatic fields that cause hum or buzz.

Shock Mount—A suspension system which mechanically isolates a microphone from its stand or boom, preventing the transfer of mechanical vibrations.

Sibilance—In a speech recording, excessive frequency components in the 5- to 10-kHz range, which are heard as an overemphasis of "s" and "sh" sounds.

Signal—A varying electrical voltage that represents information, such as a sound.

Signal Path—The path a signal travels from the input to the output in a piece of audio equipment.

Signal Processor—A device that is used to intentionally alter a signal in a controlled way.

Signal-to-Noise Ratio—The ratio in decibels between signal voltage and noise voltage. An audio component with a high signal-to-noise ratio has little background noise accompanying the signal; a component with a low signal-to-noise ratio is noisy.

Sine Wave—A wave following the equation $y = \sin x$, where x is degrees and y is voltage or sound pressure level. The waveform of a single frequency. The waveform of a tone without harmonics.

Single-D Microphone—A directional microphone having a single distance between its front and rear sound entries; such a microphone has proximity effect.

Single-Ended—This refers to an unbalanced line. Also, a single-ended noise reduction system is one that works only during tape playback (unlike Dolby or dbx, which work both during recording and playback).

Slap, Slapback—An echo following the original sound by about 50 to 200 milliseconds, sometimes with multiple repetitions.

Slate—At the beginning of a recording, a recorded announcement of the name of the tune and its take number. The term is derived from the slate used in the motion-picture industry to identify the production and take number being filmed.

SMPTE Time Code—*See also* Time Code. SMPTE is an abbreviation for the *Society of Motion Picture and Television Engineers,* which developed the time code.

Snake—A multipair or multichannel microphone cable. Also, a multipair microphone cable attached to a connector junction box.

Solo—On an input module in a mixing console, a switch that lets you monitor that particular input signal by itself. The switch routes only that input signal to the monitor power amplifier.

Sound—Longitudinal vibrations in a medium in the frequency range 20 Hz to 20,000 Hz.

Sound Generator—A synthesizer without a keyboard, containing several different timbres or voices, which are triggered or played by certain MIDI signals from a computer sequencer program, or by an external keyboard.

Sound-Pressure Level (SPL)—The acoustic pressure of a sound wave, measured in decibels above the threshold of hearing. The higher the SPL of a sound, the louder it is. dB SPL = 20 log (P/P_{ref}), where P = the measured acoustic pressure and P_{ref} = 0.0002 dyne/cm^2.

Sound Wave—The periodic variations in sound pressure radiating from a sound source.

Spaced Pair—A stereo microphone technique that uses two identical microphones horizontally spaced several feet apart, usually aiming straight ahead toward the sound source.

Speaker—*See* Loudspeaker.

Special Effects—*See* Signal Processor.

Spectrum—The output vs. frequency of a sound source, including the fundamental frequency and overtones.

SPL—*See* Sound Pressure Level.

Splice—To join the ends of two lengths of magnetic tape or leader tape with an adhesive tape. Also, a splice is the taped joint between two lengths of magnetic tape or leader tape.

Splicing Block—*See* Editing Block.

Splitter—A transformer or circuit used to divide a microphone signal into two or more identical signals to feed different sound systems.

Spot Microphone—A close-placed microphone that is mixed with more-distant microphones to add presence or to improve the balance.

Stacking Tracks—The process of recording several performances of a musical part on different tracks, so that the best segments of each performance can be played in sequence during mixdown.

Standing Wave—An apparently stationary waveform that is created by multiple reflections between opposite room surfaces. At certain points along the standing wave, the direct and reflected waves cancel, and, at other points, the waves add together or reinforce each other.

Step-Time Recording—Recording notes into a sequencer one at a time, without regard to tempo, for later playback at a normal tempo.

Stereo, Stereophonic—An audio recording and reproduction system with correlated information between two channels (usually discrete), and meant to be heard over two or more loudspeakers to give the illusion of sound-source localization and depth.

Stereo Bar, Stereo Microphone Adapter—A microphone stand adapter that mounts two microphones on a single stand for convenient stereo miking.

Stereo Imaging—The ability of a stereo recording or reproduction system to form clearly defined audio images at various locations between a stereo pair of loudspeakers.

Stereo Microphone—A microphone containing two microphone capsules in a single housing for convenient stereo recording. The capsules usually are coincident.

Studio—A room used or designed for sound recording.

Submaster—A master volume control for an output bus. Also, a recorded tape that is used to form a master tape.

Submix—A small preset mix within a larger mix, such as a drum mix, keyboard mix, vocal mix, etc. Also a cue mix, monitor mix, or effects mix.

Submixer—A stand-alone mixer, or a smaller section within a large mixing console that is used to set up a submix, a cue mix, an effects mix, or a monitor mix.

Supercardioid Microphone—A unidirectional microphone that attenuates side-arriving sounds by 8.7 dB, attenuates rear-arriving sounds by 11.4 dB, and has two nulls of maximum sound rejection at 125° off-axis.

Supply Reel—*See* Feed Reel.

Sustain—The portion of the envelope of a note in which the level is constant. Also, the ability of a note to continue without noticeably decaying, often aided by compression.

Sweetening—The addition of strings, brass, chorus, etc., to a previously recorded tape of the basic rhythm tracks.

Sync, Synchronous Recording—Process of using a record head as a playback head during an overdub session to keep the overdubbed parts in synchronization with the prerecorded tracks.

Synthesizer—A musical instrument (usually with a piano-style keyboard) that creates sounds electronically, and which allows control of the sound parameters to simulate a variety of conventional or unique instruments.

Tail-Out—Refers to a reel of tape that is wound with the end of the program toward the outside of the reel. Tape stored tail-out is less likely to have audible print-through.

Talkback—An intercom in the mixing console that is used by the engineer and producer to talk to the musicians in the studio.

Take—A recorded performance of a song. Usually, several takes are done of the same song, and the best one—or the best parts of several—become the final product.

Take Sheet—A list of take numbers for each song, plus comments on each take.

Take-Up Reel—The right-side reel on a tape recorder that winds up the tape as it is playing or recording.

Tape—*See* Magnetic Recording Tape.

Tape Loop—An endless loop formed from a length of recording tape spliced end-to-end, and which is used for continuous repetition of several seconds of recorded signal.

Tape Recorder—A device that converts an electrical audio signal into a magnetic audio signal on magnetic tape, and vice versa. A tape recorder includes a transport to move the tape, some electronics, and the heads.

Tape Sync—A signal recorded on a tape track, used to synchronize a sequencer to a tape recorder. Tape sync permits the synchronized transfer of multitrack recordings from computer memory to tape.

3-Pin Connector—A 3-pin professional audio connector used for balanced signals. Pin 1 is connected to the cable shield, pin 2 is usually connected to the signal hot lead, and pin 3 usually connects to the signal return lead. *See also* XLR-Type Connector.

3:1 Rule—The rule in audio applications which states that when multiple microphones are mixed to the same channel, the distance between microphones should be at least three times the distance that each microphone is from its sound source in order to prevent audible phase interference.

Threshold—In a compressor or limiter, the input level above which compression or limiting takes place. In an expander, the input level below which expansion takes place.

Tie—To connect electrically; say, by soldering a wire between two points in a circuit.

Tight—Having very little leakage or room reflections in the sound pickup. Also, referring to well-synchronized playing of musical instruments.

Timbre—The subjective impression of spectrum and envelope. The quality of a sound that allows us to differentiate it from other sounds. For example, if you hear a trumpet, piano, and a drum, each has a different timbre or tone quality that identifies it as a particular instrument.

Time Code—A modulated 1200-Hz square-wave signal used to synchronize two or more tape transports.

Tonal Balance—The balance or volume relationships among different regions of the frequency spectrum, such as bass, mid-bass, midrange, upper midrange, and highs.

Track—A path on magnetic tape containing a single channel of audio.

Transducer—A device that converts energy from one form to another, such as a microphone or loudspeaker.

Transformer—An electronic component containing two magnetically coupled coils of wire. The input signal is transferred magnetically to the output, without a direct connection between input and output.

Transient—A relatively high-amplitude, rapidly decaying, peak-signal level.

Transient Response—The ability of an audio component (usually a microphone or loudspeaker) to accurately follow a transient.

Transport—The mechanical system in a tape recorder that moves tape past the heads. A transport controls tape motion during recording, playback, fast forward, and rewind.

Trim—A control for the fine adjustment of level, as in a bus-trim control. Also, a control that adjusts the gain of a microphone preamplifier to accommodate various signal levels.

Tweeter—A high-frequency loudspeaker.

Unbalanced Line—An audio cable having one conductor surrounded by a shield that carries the return signal. The shield is at ground potential.

Unidirectional Microphone—A microphone that is most sensitive to sounds arriving from one direction—in front of the microphone. Some examples are the cardioid, supercardioid, and hypercardioid.

VU Meter—A voltmeter with a specified transient response, and calibrated in VU or volume units, that is used to show the relative volume of various audio signals, and used to set recording level.

Waveform—A graph of a signal's sound pressure (or voltage) vs. time of a signal. The waveform of a pure tone is a sine wave.

Wavelength—The physical length between corresponding points of successive waves. Low frequencies have long wavelengths; high frequencies have short wavelengths.

Weber—A unit of magnetic flux.

Weighted—This refers to a measurement made through a filter with a certain specified frequency response. An A-weighted measurement is taken through a filter that simulates the frequency response of the human ear.

Windscreen—*See* Pop Filter.

Woofer—A low-frequency loudspeaker.

Wow—A slow periodic variation in tape speed.

XLR-Type Connector—The part number of an ITT Cannon device that has become the popular definition for a 3-pin professional audio connector. *See also* Three-Pin Connector.

X-Y—*See* Coincident Pair.

Y-Adapter—A cable that connects two cables in parallel in order to feed one signal to two destinations.

Index

A

B

C

D

E

F

G

H

I

J

K

L

M

N

O

P

Q

R

S

T

U

V

W

X

Y

Z

How to Build Speaker Enclosures

Badmaieff and Davis

A practical guide to the whys and hows of constructing high quality, top performance speaker enclosures. A wooden box alone is not a speaker enclosure—size, baffling, sound insulation, speaker characteristics, and crossover points must all be carefully considered.

The book contains many detailed drawings and instructions for building the various basic types of enclosures, including the infinite-baffle, the bass-reflex, and the horn-projector types, as well as different combinations of these. This practical book covers both the advantages and disadvantages of each enclosure type and includes a discussion of speaker drivers, crossover networks, and hints on the techniques of construction and testing.

Topics covered include:

- Speaker Enclosures
- Drivers for Enclosures
- Infinite Baffles
- Bass-Reflex or Phase-Inversion Enclosures
- Horn Enclosures
- Combination Enclosures
- Crossover Networks
- Construction and Testing Techniques

144 Pages, 5½ x 8½, Softbound
ISBN: 0-672-20520-3
No. 20520, $6.95

Introduction to Professional Recording Techniques

Bruce Bartlett,
The John Woram Audio Series

This all-inclusive introduction to the equipment and techniques for state-of-the-art recording—whether in residences or professional studios or on location—offers a wealth of valuable information on topics not found in other books on audio recording.

Geared primarily for the audio hobbyist or aspiring professional, this book delivers a comprehensive discussion of recording engineering and production techniques, including special coverage of microphones and microphone techniques, sampling, sequencing, and MIDI. It provides up-to-date coverage of monitoring, special effects, hum prevention, and spoken-word recording, as well as special sections on recognizing good sound and troubleshooting bad sound.

Topics covered include:

- The Recording and Reproduction Chain
- Simple Home Recording
- Setting Up the Studio
- Microphones and Microphone Techniques
- Control-Room Techniques
- On-Location Recording
- Judging the Recording
- Appendices: dB or not dB, Introduction to SMPTE Time Code, and Further Education

416 Pages, 7½ x 9¾, Softbound
ISBN: 0-672-22574-3
No. 22574, $24.95

John D. Lenk's Troubleshooting & Repair of Audio Equipment

John D. Lenk

This manual provides the most up-to-date data available and a simplified approach to practical troubleshooting and repair of major audio devices. It will enable both the beginning and the intermediate level technician or hobbyist to apply tips and tricks to any specific equipment.

This book also includes such time-saving hints as circuit-by-circuit troubleshooting based on failure or trouble symptoms, universal step-by-step procedures, and actual procedures recommended by manufacturers' service personnel.

Topics covered include:

- Introduction to Modern Audio Equipment Troubleshooting
- Troubleshooting and Repair of Amplifiers and Loudspeakers
- Troubleshooting and Repair of Linear-Tracking Turntables
- Troubleshooting and Repair of Audio Cassette Decks
- Troubleshooting and Repair of AM/FM Stereo Tuners
- Troubleshooting and Repair of CD Players

208 Pages, 8½ x 11, Softbound
ISBN: 0-672-22517-4
No. 22517, $21.95

Modern Recording Techniqu Third Edition

David Miles Huber and Robert A. Runs

Recording engineers, technicians and audio engineering students appreciate this updated version the best-selling *Modern Recordi Techniques.* It has been comple updated with new information o state-of-the-art audio topics inclu ing digital audio, random access audio, and the use of digital tec nologies in audio production.

The book provides a basis for i ligence and understanding of recording technology, allowing t reader to get a feel for the enti scope of procedures. The book's comprehensive coverage makes an ideal reference for the practi or aspiring recording engineer.

Topics covered include:

- Introduction
- Sound and Hearing
- Studio Acoustics
- Microphones: Design and Technique
- The Analog Audio Tape Recorder
- Digital Technology
- MIDI and Electronic Musical Instrument Technology
- Synchronization
- The Amplifier
- The Audio Production Cons
- Signal Processors
- Noise Reduction Devices
- Monitor Speakers
- Product Manufacture
- Studio Session Procedures
- Tomorrow's Industry: Just Around the Corner

400 Pages, 7½ x 9¾, Softbour
ISBN: 0-672-22682-0
No. 22682, $26.95

Visit your local book retailer or call
800-428-SAMS.